AF553341

HUMAN TWINS

HUMAN TWINS

By

Dr. Prafulla Kumar Mohanty

&

Dr. Sanjaya Narayan Otta

DISCOVERY PUBLISHING HOUSE PVT. LTD.

NEW DELHI-110 002

First Published-2010

ISBN 978-81-8356-650-6

Published by:

DISCOVERY PUBLISHING HOUSE PVT. LTD.

4831/24, Ansari Road, Prahlad Street
Darya Ganj, New Delhi-110002 (India)
Phone: 23279245, 43764432 • Fax: 91-11-23253475
E-mail: parul.wasan@gmail.com
info@discoverypublishinggroup.com
Website: www.discoverypublishinggroup.com

Printed at:

Sachin Printers
Delhi

Preface

The aim in presenting this book to the students at undergraduate and post-graduate levels of various Indian universities is to provide a rich but concise matter pertaining to human twins.

The subject matter has been divided into seven chapters. Sincere efforts have been poured in to provide a simple and lucid account of each chapter. The diagrams are specially designed for clarity and simplicity on twinning.

We are pleased to record thanks to our colleagues at various institutions and universities all over the country who have immensely helped us by their valuable suggestions in preparing this work.

Suggestions for the improvement of the book are not only welcome but greatly appreciated.

Prafulla K. Mohanty
Sanjaya Narayan Otta

Acknowledgements

We owe a deep sense of gratitude to the Head of Post -Graduate Department of Zoology, Utkal University for permitting to carry out the project related to human twins and his encouragement during preparation of this manuscript.

We express our sincere thanks to our colleagues and friends, for their inspiration and intellectual stimulus.

We extend sincere thanks to the Director, F.P.B. for constructive suggestions on collection of finger prints. Thanks are also due to the Director, S.F.S.L. for valuable information on handwriting.

We extend thanks to Commissioner-cum-Secretary, Department of Information and Public Relation, Government of Orissa, Bhubaneswar for his sincere cooperation in providing information related to the geography of Orissa. It gives us genuine pleasure to express indebtedness to Heads and teachers of different schools and colleges of Orissa for their ungrudging cooperation and all the parents of twins for their assistance and cooperation from time to time for providing information on twins.

We express our gratefulness and appreciation to Mr. Sankar Mohanty, City Cartographer, Nayapalli, Brit Colony, Bhubaneswar for preparing the drawings, sketches, figures and map in a meticulous way.

We extend our heartfelt thanks to their parents, wives, brothers, sisters and children for their enthusiastic support during preparation of this book.

Prafulla K. Mohanty
Sanjaya Narayan Otta

Contents

1
Introduction

Twin [O.E. *getwinn* (noun)-twin; *twinn* (adj.)- double] is defined as one of two people or animals born at one birth or a person or an animal or thing very like or closely associated with another. Twins have always been a source of interest and curiosity to human beings and a matter of investigation for the scientist. Animals such as human beings usually give birth to a single individual. This phenomenon is known as "monotocous" whereas animals such as tigers, lions, cats and dogs normally produce more than one young in a single birth. These animals are "polytocous" in nature. This phenomenon, when many individuals are born is generally called superfoetation or superfecundation and when two individuals are born simultaneously, the individuals are called twins.

General Account of Development

Most of the metatherian (marsupials) and eutherian (carnivores, rodents and insectivores) mammals are polyovulatory and shed multiple ova in single ovarian cycle. Multiple births in these species are the basic principle of birth and each litter is fraternal or dissimilar to look at. Some species (e.g., marmosets) are diovulatory and discharge two ova in each ovarian cycle. The resulting twins are obviously fraternal and are produced by fertilization with two different sperms. Most primates including man in general are monovulatory and shed a single ovum in each cycle. In this case the birth of a single offspring is the rule. Twinning and multiple births are accidental and sporadic in these species. This is influenced by genetical and environmental factors. When a mother gives birth to two young individuals in single pregnancy, it is known as the twinning. Twinning in man is a hereditary trait and occurs at a frequency of about one in eighty births (Datta, 2004). This may be identical (monozygotic) or non-identical (dizygotic) which is difficult to study in case of human beings. The armadillo (*Dasypus*) habitually produces a litter of four (quadruplets) in which the members of any given litters are usually strikingly alike (Fig. 1.1). In armadillos, only one egg (a typical mammalian egg) is released from the ovary at each breeding cycle. It is believed that division of early embryo forms four independent embryos, which is due to long rate of metabolism. It is further discovered that in the early pre-twinning stages, the embryos were found to establish the connection with the uterine wall very slowly (Balinsky, 2004). This delay results in an arrest of development, with the subsequent twinning. But it is not well known, whether similar situation exists in case of identical twins, triplets and other multiple births in humans. This is probable that the condition of twinning in human beings (Fig. 1.2) is essentially the same as in armadillo.

Siamese twins are of the same origin as identical twins but they are occasionally born physically in conjoint form. The extent of the union varies from a superficial one, permitting their ready separation to a union so deep seated that the life may not be possible. Thus, the Siamese twins are the products of the incomplete separation of a single cleaved fertilized egg at about 15 days or more after zygote formation resulting into conjoined twins or double monsters. In certain cases these have been successfully separated by surgery. The original Siamese twins,

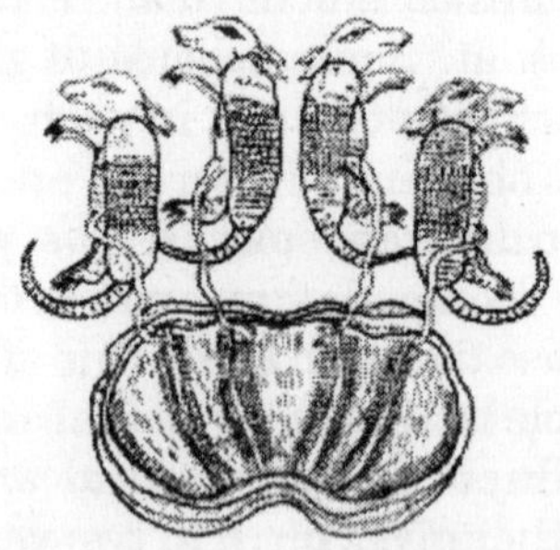

Fig. 1.1. Identical quadruplets of *Dasypus* (Armadillo).

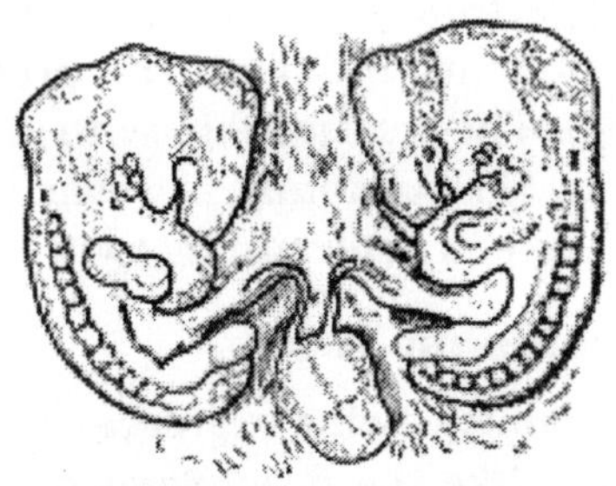

Fig. 1.2. The human embryos of almost six weeks old.

Chang and Eng, born in 1811 of Chinese parents in Siam (Thailand) were joined by a ligament from breastbone to navel. While living in retirement in North Carolina after many years of self-support by exhibiting themselves, Chang and Eng after attaining adulthood in United States each married and fathered several children.

Fraternal twins or non-identical twins or bioval twins or dizygotic twins are of the same age and are no more alike than ordinary brothers or sisters. Since the gametes from the same parent in general have different sets of genes, fraternal twins have different genotypes. The genotypes of such twins need not be any more or any less different than the genotypes of siblings (brothers and sisters) who are not twins. Fraternal twins are, therefore, siblings, born simultaneously. The fraternal twins or non-identical twins may be of the same sex or of the opposite sex. Non-identical twins arise from the simultaneous release of two separate eggs from the ovaries of the mother each fertilized by a separate sperm cell. Therefore, these twins are referred to as dizygotic twins (two eggs).

The production of twins is assumed to depend exclusively on the ability of female to produce more than one ovum for simultaneous conception. Twinning is believed to be based on a recessive genetic factor. It is unclear whether the factor for twinning follows a single factor or multiple factor types of inheritance. Nevertheless, the factor is known to be transmitted by the female as well as the male, although with different frequencies, both for monozygotic or dizygotic twins. The principal maternal factors like physical characteristics, age, menstruation and fertility influence the process of multiple births (Datta, 2004). The relationship with maternal age has some curious aspects. Most certainly, however, age of the mother plays an important role in determining twin births.

History and Records on Twins

In the human community, twins represent a biological rarity. In Indian mythology, like the *Ramayana* twin brothers Lava and Kusha are described who were the twin sons of the Lord Rama. In ancient times, various superstitious beliefs grew up around double monsters. Some people are of the concept that they are hybrid offsprings of man and other mammals. As a matter of fact, there is no evidence of the origin of a hybrid between man and any other mammal. In Hindu society there is a superstitious belief that twins are resulted due to consumption of conjoined fruits or vegetables. This misconception still exists today in our society quite firmly, for which people avoid taking conjoined fruits or vegetables. Siamese twins, also called conjoined twins, are physically joined and often share some organs. Fusion is typically along the trunk of the body or at the front, side or back of the head. In case of symmetrical conjoined twins, both children are relatively normal except for the areas of fusion. In some cases, the individuals are successfully separated by surgery.

In case of asymmetrical *Siamese* twins one is fairly normal called the host, or autosite, but the other, called the parasite, is severely underdeveloped, often tiny and dependent on its host for nutrition. Such a parasite may often be surgically separated from the host in order to save the host. The term Siamese twins originally referred to Chang and Eng born to Chinese parents in Siam (now Thailand), who were joined by a ligament from breast bone to navel. Another pair of twins, made famous in France were Ritta and Christina, a Sardinian born pair who lived only for five months but were exhibited in Paris in 1829. Ritta and Christina had separate heads, arms and vertebrae but shared a body below the navel and had two legs.

The conjoined twins Anjali and Geetanjali became news for being living examples of separated Siamese twins. Dr Y Nayudamma had taken up the case at the Guntur Government German Hospital (GGH), Guntur and successfully parted the twins in 1993. Their father, K. Sivarageswara Rao is a lorry driver. The kids were only three at the time of partition. Almost 16 years have passed by since the surgery and now they are leading a comfortable life (Fig. 1.3).

Fig. 1.3. Separated conjoined twins. Anjali and Geetanjali.

In 1997, Carson successfully separated craniopagi twins, 11 month-old Luka and Joseph Bonda, in a 28 hour operation in South Africa.

The conjoined twins Ganga and Jamuna, 34, of West Bengal, earn their living as circus exhibits. Believed to be the oldest conjoined twins in India, they share a urinary tract and kidneys. They have even conceived a child (Fig. 1.4).

Fig. 1.4. Conjoined twins Ganga and Jamun of West Bengal

The conjoined twins Mohammed Ibrahim and Ahmed Ibrahim of Egypt were separated in US in a 34-hour surgery in October, 2003 (Fig. 1.5)

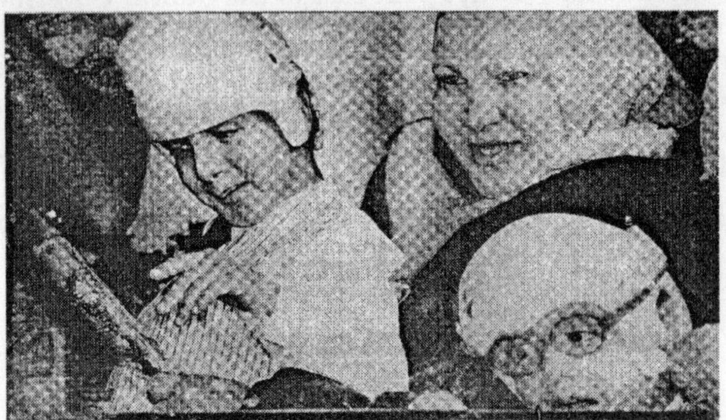

Fig. 1.5. Separated conjoined twins Mohammed Ibrahim and Ibrahim of Egypt.

The fastest operation to separate conjoined twins was in London in 2003. Zainab and Jannat Rahman, who shared a liver, were separated in 45 minutes.

In 2005 an unprecedented, 97-hour mammoth separation surgery on fused head to head Nepalese twin girls Ganga and Jamuna Shrestha in Singapore leads to their immortality. Ganga is comatose. Her brain became infected during surgery and doctors fear she will remain retarded. Jamuna has progressed well, speaks a few words and is able to interact, although she cannot walk and her skull is vulnerable (Fig. 1.6).

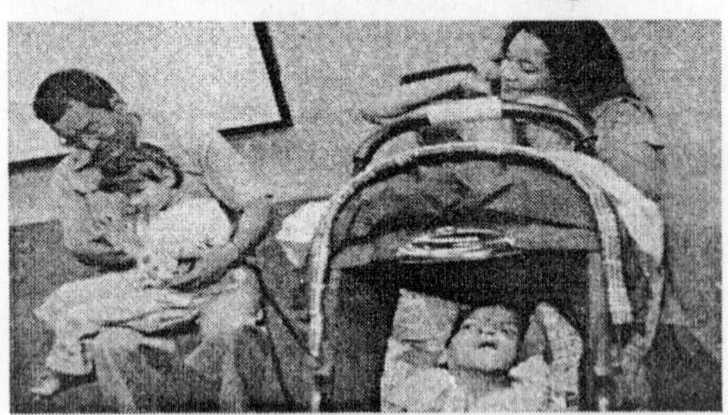

Fig. 1.6. Separated Nepalese conjoined Twin girls Ganga and Jamuna.

Siamese twins Saba and Farah Shakeel have been together for 10 years, but they may finally be able to lead separate lives if they undergo surgery after undergoing certain tests. The conjoined twins have come to the capital from Patna. The twins are conjoined at the head. They are craniopagus. Craniopagus are rare and the usual incidence of conjoined twins is one in two million births with most of them surviving long after birth. The girls do have separate brains, but they share major drainage vessels. Also, one of the girls has both the kidneys (Fig. 1.7).

Fig. 1.7. Conjoined twins Farah and Sabah of Patna.

The-two year old Sona and Mona hardly realize the complexity of their existence and enjoy the togetherness. They share two legs, kidneys, a liver, bladder, intestines and genitalia. Abandoned by their parents, the twins were later adopted by Colonel B.S. Mann and his wife Chimmney (Fig. 1.8).

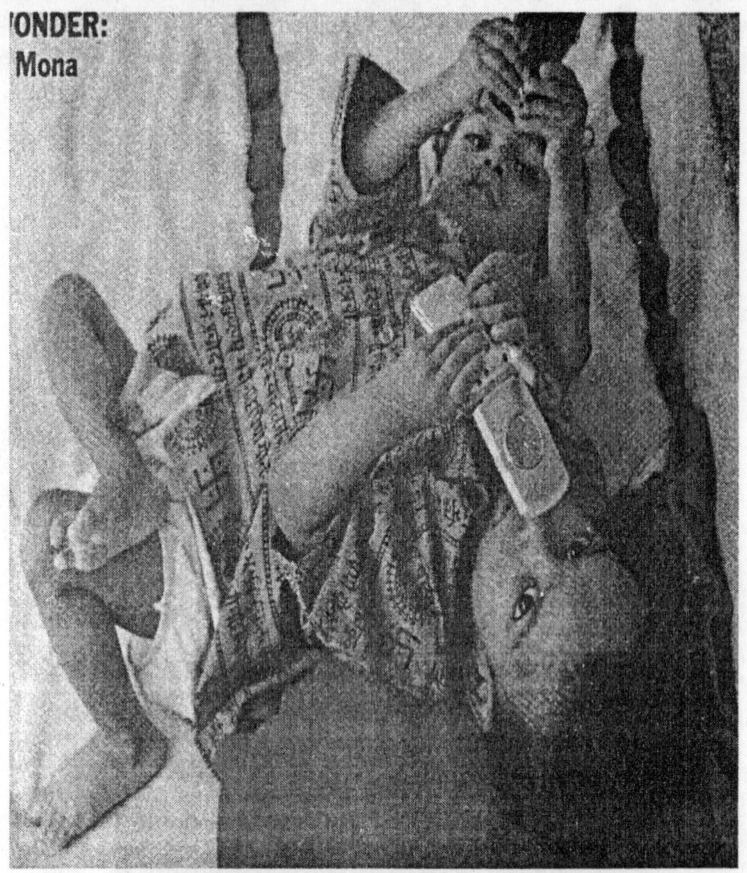

Fig. 1.8. Conjoined twins Sona and Mona

At 28, Iranian sisters Ladan and Laleh Bijani were the oldest conjoined twins to be operated upon. The surgery was a failure.

Lori and Reba Schappell, 44, of US, are the oldest craniopagi twins alive. They are joined at the top of the head (Unnithan and Ghosh, 2005).

Conjoined female twins Rejina and Reneta (Fig. 1.9) of Mexico have been separated successfully at Carlifornia in 2006. It took more than 12 hours to complete the operation.

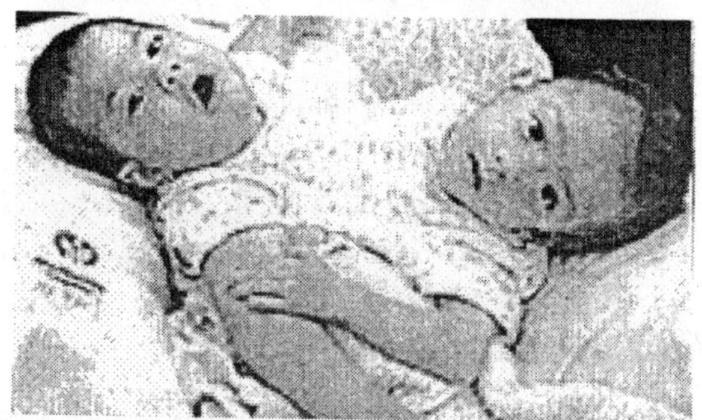

Fig. 1.9. Conjoined female twins Rejina and Reneta of Mexico.

In 2007, ischiopagus twins (Fig. 1.10) were in news as they were under observation at Ahmedabad civil hospital.

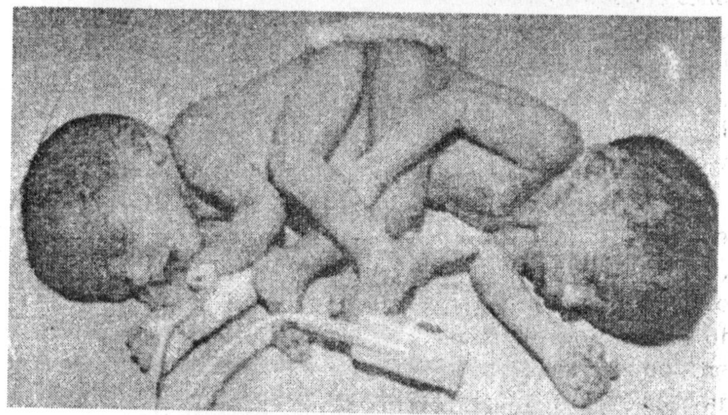

Fig. 1.10. Conjoined ischiopagus twins Twins at Ahamadabad civil hospital

Two-year old female merocephalic conjoined twin, Laxmi (Fig. 1.11) of Bihar was in the news in 2007, as she had undergone an operation for a period of 40 hrs in order to make her physically normal. She was operated and kept in a ventilator for observation at a local hospital, Sparsh of Bengaluru.

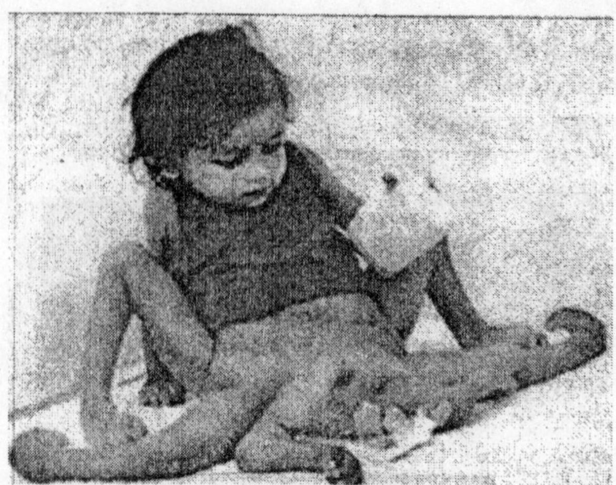

Fig. 1.11. Merocephalic conjoined. Twin Laxmi of Bihar.

In 2009, the birth of the female conjoined twins, dicephalus dibrachius dipus (Fig. 1.12) were under supervision at Fabela Memorial Hospital, Manila, Philippines.

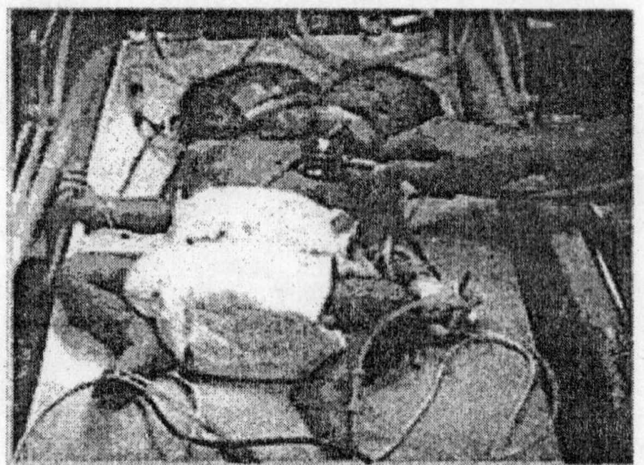

Fig. 1.12. Dicephalus dibrachius dipus Conjoined twins at Fabela Memoria Hospital, Manila, Philippines.

The records on twins are interesting and significant. The following historical data on twins have been recorded in *Guiness Book of World Record 2000 and 2002 and Limca Book of Records* (Table. 1.1).

(1) Heaviest Twins at Birth: The world's heaviest twins were born to J.P. Haskin (USA) on February 20, 1924 (Field *Masson, Robins and Broadley*, 2002). They had a combined weight of 27 lb 12 oz (12.6 kg).

(2) Heaviest Twins: Billey Leon and Benny Loyd (Both USA) were normal in size until the age of six (Filed, *Masson, Robins and Bradley*, 2002). In November, 1978, just before their 32nd birth day, Billey and Benny weighed 743lb (337 kg) and 723 lb (323 kg) respectively, and both had waist measuring 84 inches (2.13m). They became professional tag wrestling performers. When. weighed before competition, they clocked weight of up to 769 lb (349 kg).

(3) Lightest Twins: Two sets of twins have been born with a combined weight of 30.30 oz (860g). Roshan Maralyn weighed 17.28 oz (409g) and Melanie Louise weighed 13.05 oz (370g) were born to Katrina Grang of Australia at the Royal Women's Hospital, Brisbane, Australia, on November 19, 1993. Wendy Morrison of Canada gave birth to Anne (14.81 oz or 420g) and John (15.52 oz or 440g) at Ottawa General Hospital, Ontario, Canada on January 14, 1994 (Field, *Masson, Robins and Broadley*, 2002).

(4) Shortest Living Twins: The shortest living twins are John and Grey Rice (USA) born on December 3, 1951 (Field *Masson, Robins and Bradley*, 2002). They are identical twins and both measure 2ft 10 inches (86.3 cm).

(5) Shortest Twins: The shortest twins on record (Field *Masson, Robins and Bradley*, 2002) are Matyus and Bela Matina (Hungary, later USA; 1903 –C. 1935), who were both 2ft 6 inches (76 cm.) tall.

(6) Largest Gathering: On November 12, 1999 a national holiday marking the birth of Sun Yatsen, the first president of the Republic of China – a total of 3,961 pairs of twins gathered on the square of Taipei City Hall, Taiwan to break the record for the largest gathering of twins. The twins came from as far as the UK, Germany, India and the USA and ranged in the age from one month to 88 years. National celebrities were also present, along with 37 sets of triplets and four sets of quadruplets (Field, *Masson, Robins and Bradley*, 2002).

Table 1.1. Records of Twins in the World

Sl. No.	*Record on twins*	*Description of the record*
1.	Heaviest twins at birth	Born to J.P. Haskin (USA) on Feb 20, 1924 having combined weight of 27 lb 12 oz (12.6 kg).
2.	Heaviest twins	Billey Leon and Benny Loyd of USA just before their 32nd birthday Billey and Benny weighed 743 lb(337 kg) and 723 lb (328 kg).They weighed before a wrestling competition, the clocked weights of up to 349kg.
3.	Lightest twins	One set of twins has been born to Katrina Gray of Australia at the Royal Women's Hospital, Brisbane, Australia with a combined weight of 30.30 oz (860g). Wendy Morrison of Canada gave birth to other set of Anne (14.81) oz or 420 g and John (15.52 oz or 440g) at Ottawa General Hospital Ontario, Canada on Jan 14, 1994.
4.	Shortest living twins	John and Grey Rice (USA), as identical shortest living twins born on Dec 3, 1951 and measured 2 ft 10 inches (86.3 cm).
5.	Shotest twin ever	Matyus and Bola Matina (Hungary, later USA) who were both 2 ft 6 inches(76cm) tall.
6.	Largest gathering	On Nov 12, 1999, a total of 3,961 pairs of twins gathered on the square of Taipei City Hall, Taiwan to break down the record for the largest gathering of twins.
7.	Longest time a twin has remain undiscovered	In July 1997, a foetus was discovered in the abdomen of 16-year old Hisham Ragab of Egypt bearing length 18cm and weight 2kg (4-lb 6-oz).
8.	Longest living conjoined twins	Chang and Eng Bunker from Siam born on 11 May, 1811 and died with in three hours of each other at the age of 63 on 17 Jan,1874.

Sl. No.	*Record on twins*	*Description of the record*
9.	Most successful conjoined twins of modern time	Andy Gracia, the US actor who has starred in few films was born with a twin attached to his shoulder in Cuba in 1956.
10.	Longest living two headed person	The two-headed boy of Bengal was born in 1783 and died from a Cobra bite at the age of 4.
11.	Most twins in a school	On October 1999, St. Joseph's High School for Girls in Cherganncherry in Kerala had 17 pairs of twins from Std I to Std X. All the pairs of twins except for Jobin and Jobin of class III(Junior School) were girls.
12.	First separation of conjoined twins	Dr Jabal surgically separated a set of Siamese twins born to Ms Kambli on August 11, 1962 at Bhatia Hospital, Bombay.
13.	Most twins in a row	Ujiben Barvad of Rajkot gave birth to twins for the third successive time in March 1988.
14.	First gift twins	By using first gamete intrafallopian transfer technology (GIFT), twins were born to Renuka and both of them were males.
15.	Twin brothers securing same scoring	In 1960, twin brothers of Darjeeling had scored equal number and were topper in matric. In 2005, twin brothers of 24-pragana (West Bengal) have secured 577 marks and get first position.
16.	First twins born to different women	First, biological mother Amy Bernaba gave birth to Lauren, weighing 7 lb 10 oz, then, half an hour later, surrogate mum Torry Keay delivered 7 lb 30 oz Hannah. The girls are the world's first twins to be born on the same day to two different women.

Source : Limca Book of Records 2000 and *Guiness World Records 2000 and* 2002.

(7) Longest Time a Twin has Remained Undiscovered: It is reported (Kynaston, 2000) that in July 1997 a foetus was discovered in the abdomen of 16-year old Hishma Ragab of Egypt, who had been complaining of stomach pains. A swollen sac found pressing against his kidney turned out to be his 18cm long (7 inches) and 2kg (4-lb 6oz) identical twin. The foetus, which had been rowing inside him, had lived until 32 or 33 weeks after conception.

(8) Longest Living Conjoined Twins: Chang and Eng Bunker, the conjoined twins from Siam (now Thailand) were born on 11th May, 1811, married sisters Sarah and Adelaide Yates of Wilkes country, North Carolina, USA and fathered 22 children between them (Kynaston, 2000). The twins were separated surgically for their independent survival. They died with in three hours of each other at the age of 63 on 17th January, 1874.

(9) Most Successful Conjoined Twin of Modern Times: Andy Gracia, the US actor who has starred in films such as The God Father part 3 (USA, 1990) and Things To Do In Denver When You Are Dead (USA, 1995), was born with a twin attached to his shoulder in Cuba in 1956. The twin was no bigger than a tennis ball and was removed by surgeons soon after birth (Kynaston, 2000).

(10) Longest Living Two Headed Person: The two-headed boy of Bengal was born in 1783 and died from a Cobra bite at the age of four. His two heads, each of which had its own brain, were of the same size and were covered with black hair at their junction (Kynaston, 2000).

(11) Most Twins in a School: The information as to the most twins in a school (Ghose, 2000) is that in October 1999 is St. Joseph's High School for Girls in Chengnnacherry in Kerala had 17 pairs of twins from Std. I to X. The maximum number of twins, four, were in class VII. The eldest was 14 and the youngest was five years old. All the pairs of twins except Jobin and Jobin of Class III (Junior School) were girls.

(12) First Separation of Conjoined Twins: It is reported

(Ghose, 2000) that Dr Dalal Surgically separated a set of Siamese twins born to Ms Kambli on August 11, 1962 at Bhatia Hospital, Bombay. The twins were perfectly normal.

(13) Most Twins in a Row: Ujiben Barvad of Rajkot gave birth to twins for the third successive time in March 1988 (Ghose, 2000).

(14) First Gift Twins: The first gamete intra-fallopian transfer (GIFT) technology was used for the birth of twin and both the members of twin were males.The twins were born to Renuka (Ghose, 2000).

(15) Twin Brothers Securing same Scoring: In 1960, twin brothers (Sri Asim Dasgupta and Sri Atish Dasgupta) of Darjeeling district (North Bengal) had scored equal number in the matriculation examination and were joint topper. Another pair of twin brothers of Gaighat of 24-Pragana (West Bengal) has scored 577 marks in the matriculation examination after a long span of 45 years, i.e., in 2005 and have occupied first division. These two brothers are Sri Rajat Sarkar and Sri Rakesh Sarkar by name. Rajat and Rakesh were born on 08th January, 1990 and Rajat is elder to Rakesh by three hours only. These two brothers were fond of taking rice and dal. The number of their family members was six (Anonymous, 2005).

(16) First Twins Born to Different Women: The girls Lauren and Hannah are the world's first twins to be born on the same day to two different women. First, biological mother Amy Bernaba gave birth to Lauren, weighing 7lb 10 oz, then, half an hour later, surrogate mum Torrey Keay delivered 7 lb 30 oz Hannah. The double pregnancy happened after Bernaba and husband George had undergone *in vitro* fertilization (IVF) treatment for 12 years. The doctor had implanted the eggs fertilized by Bernaba's sperm into her womb and also into surrogate Keay's. Keay became pregnant, and against all the odds so did Bernaba. Both women had straight forward pregnancies before giving birth on May 27, 2007 in neighbouring rooms in a Los Angeles Hospital. Although the twins are not identical because they came from separate eggs,they look very alike.

Review of Literature

Twinning is an interesting aspect of embryology, which has attracted the attention of various cross-sections of scientists such as biologists, zoologists, embryologists, endocrinologists, physiologists, ethologists, geneticists and environmentalists. The meaning of twin and Siamese twin of human has been explained by Goetz (1979). Unnithan and Ghosh (2005) reported on various types of conjoined twins of the world of present era. Ghosh (2000) has compiled various records on twinning in *Limca Book of Records*. Various world records on twins have been edited by Kynaston (2000) and Field *et al.* (2002) in *Guiness World Records 2000 and 2002* respectively.

For the development of science of genetics, Muller and Altenburg (1919), Watson and Crick (1953a), Wagner (1955), Jacob and Monod (1961), Sturtevant (1965), Atherly *et al.*(1999) and Klug and Cummings (2003) have rendered service to a great extent and expressed their views on various aspects, relating to heredity and hereditary units or genes which are formed of DNA. The relative frequencies of various kinds of genes in a large and randomly mating sexual panmictic population tend to remain constant from generation to generation. In the absence of mutation, selection and gene flow is stated by Hardy (1908). Mc Clintock (1941) and Watson and Crick (1953a) have reported the effect of mutant genes in animals. A good number of work on gene at molecular level has been experimented by Turpin and Lejeune (1969), Brown (2002) and Lewin (2002) and they have stated that genetic constituents of the members of monozygotic twins are alike. Altenburg (1957), Strickberger (1985) and Hartl and Jones (2001) have investigated on pedigrees which reveal patterns of inheritance of twinning in humans. According to Bergsma (1965), foetal abnormality of twins can be detected through certain methods like amniocentesis and karyotyping.

The effect of mutant gene in the members of identical twins has been studied by McClintock (1944). Muller (1974) has proposed for the establishment of sperm and egg banks, since sex cells can be kept alive and functional for a considerable period by freezing. Muller also contributed his work on radiation genetics and described artificial inductions of mutations in animals by X-rays. This also leads to variation in the members of monozygotic twins. Investigation on congenital malfunctions in twins has been made by Hay and Wehrung (1970). Erlenmeyar-Kimling and Jarvik (1963) and Bodmer and Cavalli-Sforza (1970) have narrated the work on intelligence of twins in relation to inheritance. Synthesis of nucleic acids in *in vitro* condition and isolation of DNA polymerase I have been described by Kornberg (1974). Micklos *et al.* (2003) have discussed about DNA replication and have studied genetic constituents of monozygotic twins. Various principles of inheritance pertaining to human twinning have been discussed by Gardner *et al.* (1991) and Snustad and Simmons (2003). According to Richard (1985), the birth of twins is hereditary which cannot be ignored. He is also of the opinion that the intelligence, sharpness and academic performance both in identical and non-identical twins depend on the environment. It is further added that intelligence quotient of identical twins reared apart also varies significantly (Schields, 1962 and Richard, 1985).

The genetic epidemiology of age-related maculopathy in various types of twins shows a strong genetic component (Meyers and Zachary, 1988; Dosso and Bovet, 1992; Meyers, 1994; Meyers *et al.* 1995). The twin study paradigm is a powerful method to study complex and heterogenous trait disorders. Differences between identical twins would presumably reflect environmental influences while differences between non-identical twins may be due to heredity, environment or both. Therefore, if alcoholism has a hereditary basis, monozygotic twin pairs should tend to be more similar to their drinking behaviour and alcohol-related problems than dizygotic twin pairs (Pickens *et al.* 1991). Various studies report the genetics of alcohol metabolism and alcoholism in the form of genetic epidemiology on twins (Kaig, 1960; Closinger *et al.*, 1981; Gurling *et al.*, 1981; Hrubec and Omenn, 1981; Koskenvuo *et al.*, 1984; Kaprio *et al.*, 1987; Romanov *et al.*, 1991; Kendler *et al.*, 1992; Mc Gue *et al.*, 1992; Heath *et al.*, 1997; Prescott and Kendler, 1999). Several twin studies have demonstrated that part of the variation in red cell indices between individuals is due to genetic factors (Lindemnn *et al.*, 1977; Whitfield and Martin, 1985; Dal Colletto *et al.*, 1993; Yokoyama and Akiyama, 1995; Evans *et al.*, 1999; Zhu *et al.*, 1999; Garner *et al.*, 2000; and David *et al.*, 2001).

Frankham *et al.* (2002) have proposed conservation of genes of different species of living organisms by storing gametes in the sperm and egg banks. According to them, gametes can be obtained as and when required and serve as donors for a considerable period, even after the death of the parents. Genetic influences on the electro-encephalogram of twins reared together and apart have been interpreted by Young *et al.* (1972) and Stassen *et al.* (1988). Bodmer and Cavalli-Sforza (1970) and Novitski (1977) have reported on the variations of the behaviour and intelligence quotient between the members of twins.

Herbert and Graham (1974) differentiated different regions of early mammalian embryo by application of different chemicals on the embryo. The role of epithelium of seminiferous tubule of man has been studied by Clermont (1963). Carlson (1994), Gilbert (2003) and Balinsky (2004) reviewed the development of human embryo in detail. The development of different stages human embryo has been extensively studied by Wolpert (2002). Development of ovary, stages of maturation of ova and fertilization have been described by Hamilton (1944), Shettles (1958), Pinkarton (1961), Crosignani and Mishell (1976), Bedford (1977) and Zuckerman and Wair (1977). Jones and Fox (1977) have reported on ultrastructure of human placenta of monozygotic and dizygotic foetus and explained the relationship between chorionic villi and endometrium layer of uterus of mother. Development of human embryos in the laboratory has been experimented and explained by Edwards and Fowler (1970). Cole and Cupps (1964), Cohen (1977) and Datta (2004) have reported on essentiality and process of reproduction in human and other living organisms. Physiology of pregnancy, prophylaxis and treatment of complications during pregnancy, inflammatory diseases of female genitalia, abnormal sexual development of girls and embryology of twinning have been studied and explained by Bodyazhina (1987). Ursprung (1967) has discussed the formation of various patterns of the development of embryo in animals. Development of human in relation to gene expression during development has been studied by Markart and Ursprung (1971). Weil (1965) has experimented on activities of spermatozoa in the seminal vesicle and suggested, how coating of spermatozoa acts as antigen of the seminal vesicle.

A good number of workers such as Gordon *et al.* (1977) and Schmidt-Nelson (1977) have undertaken the studies on physiology such as respiration, circulation, excretion, nerve conduction and biosynthesis of organic materials in human twins and other animals. Tienthoven (1968) and Chang and Hunter (1975) have stated on reproductive physiology of vertebrates and capacitation of mammalian sperms. Physiology of various organs like digestive, respiratory, circulatory, excretory, nervous and reproductive system of human has been studied in detail by Chatterjee (1988 and 1992) and Guyton and Hall (2000). Hay and Wehrung (1970), Schmidt-Nelson (1977) and Chatterjee (1988) have described the development of monozygotic twins, dizygotic twins and multiple births. According to them monozygotic twins are always of the same blood groups and tissues of same antigenic potencies whereas, dizygotic twins may be either of same or opposite sexes having separate antigenic constitution. Newman *et al.* (1937) and Myrianthopoulas (1970), the type of twinning may be hereditary. Twinning occurs once in about 75 births and about two-thirds of them are of dizygotic origin. He is of the opinion that multiple births are meant for birth of more than two, e.g., triplets, quadruplets or quintuplets or more from any of the above origin or from a combination of both of them (monozygotic twins and dizygotic twins). Langman (1969) and Mc Lachlan (1994) describe the development of human twins, their physiology and importance of twin study in medical science.

Berkowitz *et al.*(1996) and Barton *et al.* (1996) have narrated the duration of gestation decreases with the increase number of fetus. The median gestational age of triplets and quadruplets is reported by Newman *et al.*(1989) and Elliott and Radin (1992). Rothman (1970) has opined the consequence of deficiency of folic acid in twin pregnancy. The mode of delivery of twins is stated by Grant (1989). Baldwin (1991), Robertson and Neer (1988),Galea *et al.*(1982) and Mahone *et al.*(1993) have cited vascular communications between fetuses. During the neonatal period polycythemia may lead to severe hyperbilirubinemia and kernicterus (Mahone *et al.,* 1993), Landy *et al.*(1986) has explained regarding vanishing twin syndrome.

Raven (1962) has contributed in the field of coloured progressive matrices in order to find out the intelligence quotient (I.Q.) of the human beings. Importance of twin studies for the individual differences has been investigated by Segal (1990). Segal and Hershberger (1999) have conducted same analysis on cooperation and competition between the members of twin and presented a report on the evolution of behaviour of human. Psychology of the twins in the classroom has been studied by Segal and Russel (1992). Segal (1993) also narrates twin and sibling in relation to their methods of adoption. Segal *et al.* (1994) report regarding the analysis of odor identification and perception of members of twins. Behavioural aspects of twins are also recorded by Bulmer (1770), Segal (1997a) and Segal and Mac Donald (1998). Segal (1997b) has experimented on environmental influence on intelligence quotient of the Siblings within a family. Segal and Ream (1998) provide an account on decrease in grief intensity for deceased twin and non-twin relatives in connection to evolutionary perspective and individual differences. According to Gottesman and Shields (1972), and Segal and Mac Donald (1998), genes are responsible for the psychology. Mittler (1976) and Mather and Black (1984) opine that the language development in young twins is influenced by heredity and environment. Report on behaviour of living organisms has been presented by Jennings (1965), Morton (1972) and Tucker (1997). Eysenck (1990) has investigated the environmental impact on the individual and noticed differences in the personality development of the individual. Vogel *et al.* (1982) have reported on behavioural maturation of twins. Vocal fundamental frequency in twin has been experimented and described by Przybyla *et al.* (1992), Nolan and Oh (1996) and Whiteside and Rixon (2001). Richard (1985) has accounted on the intelligence of twin children of different racial groups.

Variations of the handwriting, spacing and influence of writing instrument have been described by Osborn and Albert (1929), Teltscher and Henry (1948) and Harrision (1958). Baker and Newton (1955) have studied on size and stroke of the handwriting, which is dependant on writing material, writing surface, attitude of the candidate and practice. The psychology of the members of identical twin and non-identical twin through their handwriting has been investigated and summarised by Saudek (1954) and Holt and Arthur (1965). Hardless (1970) narrated on different pattern and rhythm in handwriting. Arrangements of handwriting in twin children have been studied by Bates and Billy (1970). Singer (1969) and Mehta (1970) have investigated and analysed on variations in handwriting of human beings and twins in particular.

Various works on directories of fingerprints have been narrated by Galton (1895). Larson (1924), Bose (1927) and Scott (1951) have put forth general description of fingerprint patterns. Comparative account of fingerprints has been analysed by Brewester (1936), Browne and Brock (1953) and Bridges *et al.* (1963). Alexander (1975) has investigated on identification on thumb impression. Comparison of fingerprints, single digit system and application of fingerprint system have been described by Galton (1965). Moenssens (1975) has worked out on techniques used in study of fingerprints.

Application of various methods, research methodology, application of various principles of statistics, sampling, data analysis and development of questionnaire in research work has been elaborated by Bajpai (1960) and Paneerselvam (2004). A vivid description of statistical application in biology relating to twin has been made by Zar (2003) and Rao and Richard (2004).

The particular investigation on "Genetic study on urban and rural twins of Orissa" has been undertaken to study, analyse and interprete the various aspects such as morphological features, physiological activities,ethology and dactylography of monozygotic twins,dizygotic twins and triplets of Orissa.

For the present investigation, 103 pairs of twins and five triplets have been collected from both urban and rural areas of Orissa (Fig. 2.1). These 103 pairs of twin pairs include 52 pairs of identical twins and 51 pairs of non-identical twins. Out of these 52 pairs of identical twins (Figs. 1. 52 BLS 1.103 RGD 1), 34 pairs are males and rest 18 pairs are females. Further, out of 51 pairs of non-identical twins (Figs. 1.51. BDK 1.51 SMP1), 15 pairs are males, 11 pairs are females and 25 pairs consist of one male and one female in each pair. In triplets, out of five sets, two sets include two male members and one female member in each set, third set includes one male member and two female members (out of two female members one was dead) and rest two sets comprise of three female members in each set.

In toto 103 pairs of twins and five sets of triplets have been analysed from rural and urban areas of different districts of Orissa, namely Balasore, Bhadrak, Cuttack, Ganjam, Jajpur, Jagatsingpur, Kendrapara, Keonjhar, Khurda, Mayurbhanj, Nayagarh, Puri, Rayagada and Sambalpur (Fig. 1.13).

Out of 51 pairs of non-identical twins, 26 pairs and 25 pairs of twins belong to different rural and urban areas respectively. Further, 20 pairs from 52 pairs of identical twins belong to the different rural areas and 32 pairs are from different urban areas. From 51 pairs of non-identical twins, 41 pairs have been reared together whereas 10 pairs have been reared apart. But incase of identical twins, out of 52 pairs, 43 pairs have been reared together and nine pairs have been reared apart. Out of the five sets of triplets, four sets belong to urban areas and one set is from rural area. Interestingly enough, all the five sets of triplet are reared together. The entire study was undertaken from the year 1999 to 2005.

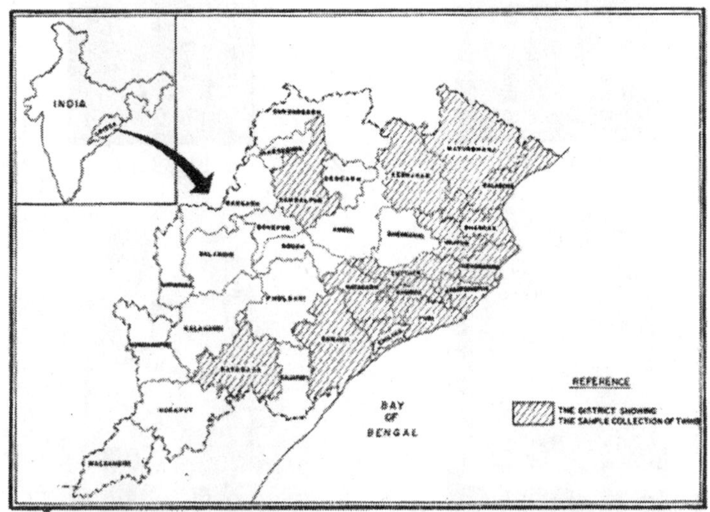

Fig. 1.13. Map of Orissa

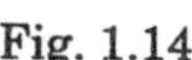

Fig. 1.14

Fig. 1.15

Non-Identical Twins

Fig. 1.16

Fig. 1.17

Fig. 1.18

Fig. 1.19

Fig. 1.20

Fig. 1.21

Non-Identical Twins

Fig. 1.22

Fig. 1.23

Fig. 1.24

Fig. 1.25

Fig. 1.26

Fig. 1.27

Non-Identical Twins

Fig. 1.28

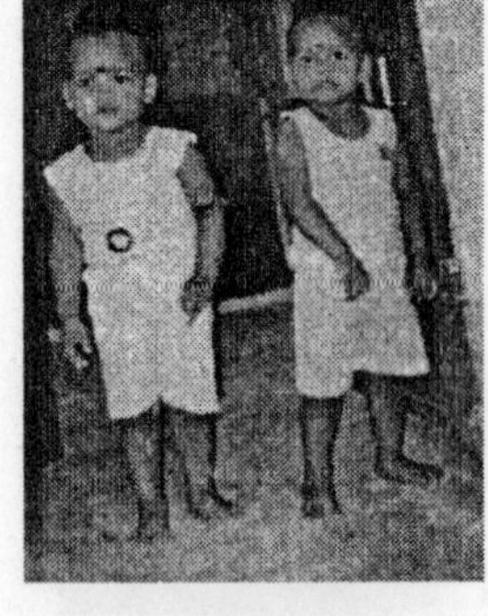

Fig. 1.29

Fig. 1.30

Fig. 1.31

Fig. 1.32

Fig. 1.33

Non-Identical Twins

Fig. 1.34

Fig. 1.35

Fig. 1.36

Fig. 1.37

Fig. 1.38

Fig. 1.39

Non-Identical Twins

Fig. 1.40

Fig. 1.41

Fig. 1.42

Fig. 1.43

Fig. 1.44

Fig. 1.45

Non-Identical Twins

Fig. 1.46

Fig. 1.47

Fig. 1.48

Fig. 1.49

Fig. 1.50

Fig. 1.51

Non-Identical Twins

Fig. 1.52

Fig. 1.53

Fig. 1.54

Fig. 1.55

Fig. 1.56

Fig. 1.57

Non-Identical Twins

Fig. 1.58

Fig. 1.59

Fig. 1.60

Fig. 1.61

Fig. 1.62

Fig. 1.63

Fig. 1.64

Non-Identical Twins

Fig. 1.65

Fig. 1.66

Fig. 1.67

Fig. 1.68

Fig. 1.69

Fig. 1.70

Non-Identical Twins

Fig. 1.71

Fig. 1.72

Fig. 1.73

Fig. 1.74

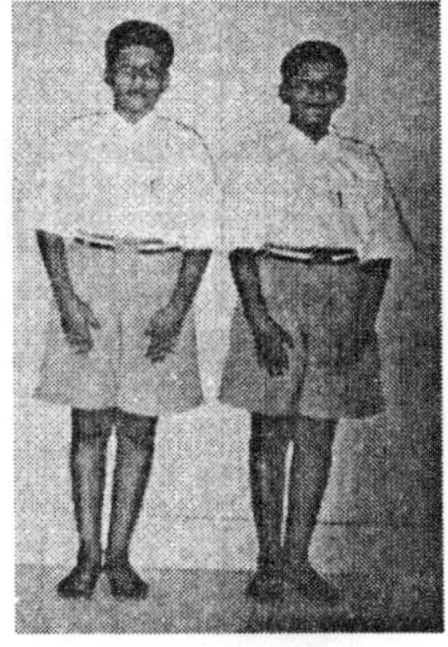

Fig. 1.75

Fig. 1.76

Non-Identical Twins

Fig. 1.77

Fig. 1.78

Fig. 1.79

Fig. 1.80

Fig. 1.82

Fig. 1.82

Non-Identical Twins

Fig. 1.83

Fig. 1.84

Fig. 1.85

Fig. 1.86

Fig. 1.87

Fig. 1.88

Non-Identical Twins

Fig. 1.89

Fig. 1.90

Fig. 1.91

Fig. 1.92

Fig. 1.93

Fig. 1.94

Non-Identical Twins

Fig. 1.95

Fig. 1.96

Fig. 1.97

Fig. 1.98

Fig. 1.99

Fig. 1.100

Non-Identical Twins

Fig. 1.101

Fig. 1.102

Fig. 1.103

Fig. 1.104

Fig. 1.105

Fig. 1.106

Non-Identical Twins

Fig. 1.107

Fig. 1.108

Fig. 1.109

Fig. 1.110

Fig. 1.111

Fig. 1.112

Non-Identical Twins

Fig. 1.113

Fig. 1.114

Fig. 1.115

Fig. 1.116

Fig. 1.65-1.116. Non-Identical Twins.

Information regarding the availability of twins pairs were collected from informers such as Principals, Headmasters, staff members of colleges, schools, relatives and friends. Accordingly, parents and twins were contacted over phones and as per their convenient timing, data were collected from their residence or from schools and colleges. During non-availability of telephone, persons were met directly and time was fixed to collect information about the twins.

2

Twinning : Conceptual Analyses

The simultaneous birth of two youngs at a single pregnancy is called twins. Twinning, common in many animals, is of two biological kinds such as the one egg (monozygotic) or identical type and the two eggs (dizygotic) or fraternal type. The latter is more usual and can be thought of simply as a litter of two. In man, psychological studies of sets of identical twins, since they are genetically identical, have provided much otherwise unobtainable information on the relative effects of genetic endowment and environment. Twinning is a phenomenon of giving birth to two offsprings at a time. In this case, two foetuses are present in the uterus. Twin birth comes under multiple births, which is only a chance of occurrence. Multiple births mean a process in which two or more than two offsprings are born at a time. But until today the conception of twins and its birth is miraculous not only in history but also in the field of developmental biology.

Embryological Aspects of Twinning

In human beings, sexes are separate and reproduction takes place by sexual method. The male reproductive system of a man consists of a pair of testes (primary sex organs) present inside the scrotum, epididymis, vasa deferentia, ejaculatory ducts, seminal vesicles, prostate glands, Bulbo-urethral glands and a penis or the copulatory organ. The female reproductive system of a human includes a pair of ovaries (*primary sex organs*), Fallopian tubes, uteri, vagina, vulva and Bartholin's glands and a pair of mammary glands. In human beings, the male and female gametes are respectively the sperms and ova.

Spermatogenesis is a process of the formation of sperms, whereas oogenesis is the formation of ova.

Formation of gametes starts at puberty that starts at the age of 9 to 11 years in girls and 11 to 13 years in boys. Fertilization is the union of two opposite types of gametes. Sexual intercourse or coitus accomplishes sperm transfer. During copulation, semen-containing millions of sperms is released into vagina (insemination). A single ejaculation of semen may contain three hundred million sperms. From the vagina the sperms travel up the uterus but only a few thousands find their way into the passage of Fallopian tubes.

The ovum is transferred through the fimbriated end of the tube by muscular contractions aided by ciliary movements of its epithelium. The area of ovum, which extrudes the polar bodies and receives sperm, is termed the animal pole. The fusion of a haploid male gamete (sperm) and a haploid female gamete (ovum) to form a diploid zygote is called fertilization (Fig.II.1). In human being, fertilization takes place mostly in the ampulla of the oviduct (Fallopian tube). About thirty hours after fertilization, the newly formed zygote divides into two cells, the blastomeres, in the upper portion of the Fallopian tube. This is the first cleavage. The next division occurs within forty hours after fertilization. The third division occurs about three days after fertilization. During these early cleavages, the young embryo slowly moves down the Fallopian tube towards the uterus.

At the end of the fourth day, the embryo reaches the uterus. It has 32 cells. This solid mass of cells is known as morula (little mulberry) as it looks like a mulberry. At the next stage of development, which produces an embryo with about 64 cells, a cavity is formed within the cell mass. This cavity is called blastocoel and the embryo is termed as blastocyst, which is composed of an outer envelope of cells, the trophoblast and inner cell mass. The trophoblast encircles the blastocoel and the inner cell mass. The latter is the precursor of the embryo. It means the inner cell mass gives rise to the embryo. The cells of the trophoblast form the placenta and foetal (embryonic) membrane. The extra embryonic or foetal membranes are the amnion, chorion, allantois and yolk sac. Amnion provides a fluid medium to the developing embryo. It prevents desiccation of the embryo and functions as a shock absorber. As the human egg is devoid of yolk, the yolk sac develops as an evolutionary process. It is very small and gradually degenerates and shrinks. The chorion and allantois take part in the formation of placenta. The human placenta is, thus, referred to as "chorioallantoic placenta". The placenta is a temporary association between the foetal and maternal tissues. It consists of foetal part and a maternal part the decidua basalis. The foetal placenta grows intimacy and invades the uterine mucosa with its chorionic villi. The degree of intimacy is so strong that eventually the blood vessels of the villi are literally bathed in mother's blood. The umbilical cord connects the foetus to the placenta.

This process of the attachment of blastocyst with the uterine wall of the mother is called implantation (Figs. 2.1 and 2.2). It occurs after seven days of fertilization. The trophoblast is destined to form the extra embryonic mesoderm and the chorionic villi embedding into the uterine mucosa, the endometrium. The blastocyst then undergoes gastrulation to produce the three primary germinal layers.

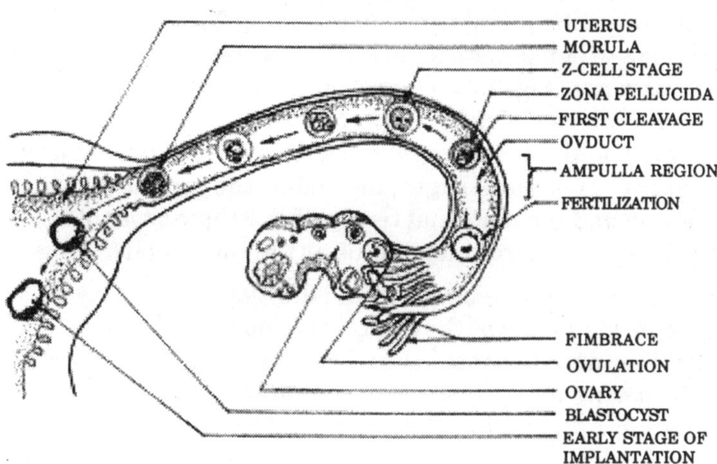

Fig. 2.1. Development of human embryo in female reproductive tract (Fertilization-implantation).

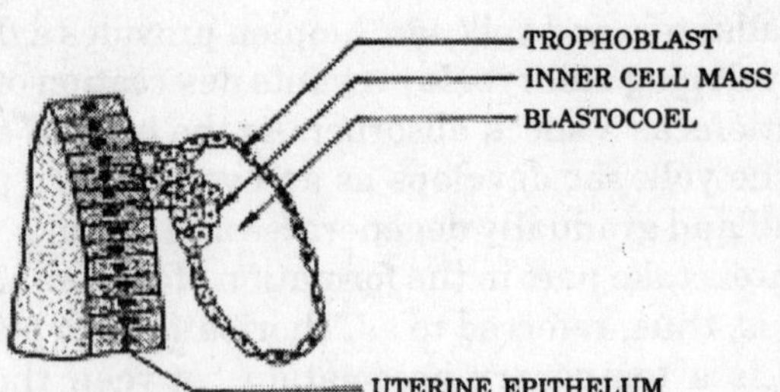

Fig. 2.2. Implantation of a human embryo on the uterine wall.

This involves cell movements (morphogenetic movement) that eventually help to attain new shape and morphology of the embryo. By three months, all the systems of the baby are formed, at least in a rudimentary way. From then on, development of the foetus, as it is now called, is primarily a matter of growth and minor structural modifications. At the time of a twin birth, the obstetrician usually makes a preliminary determination of the type based on the distribution of foetal and maternal tissues that make up the after birth. These include the amnion, chorion and placenta.

Early in development of the largely undifferentiated mass cells, an amniotic cavity is formed within which the embryo develops. The membrane surrounding the cavity becomes the amnion which becomes the inner most environment of the birth membranes (Fig. 2.3). The part of the outermost layer of the cell mass forms a chorion, which on further development becomes thinner and forms a second cell layer around the foetus. The placenta is of dual origin, formed at the juncture of the foetal chorion and the maternal tissue. These three types of tissue may individually surround a foetus at the time of birth (Fig. 2.4).

Since both monozygotic and dizygotic twins may develop as a separate embryo, they may develop with separate amnions, chorions and placentas (Fig. 2.5.iv and Fig. 2.6A). Monozygotic (MZ) twins may have a single chorion (dizygotic or DZ twins never do), and within that chorion may have either double (Fig. 3.5ii and Fig 2.6B) or single (Fig.2.5i and Fig. 2.6C) amnions. From this, it is clear that the identification of twins as monozygotic or dizygotic is not ambiguous unless there is a single chorion.

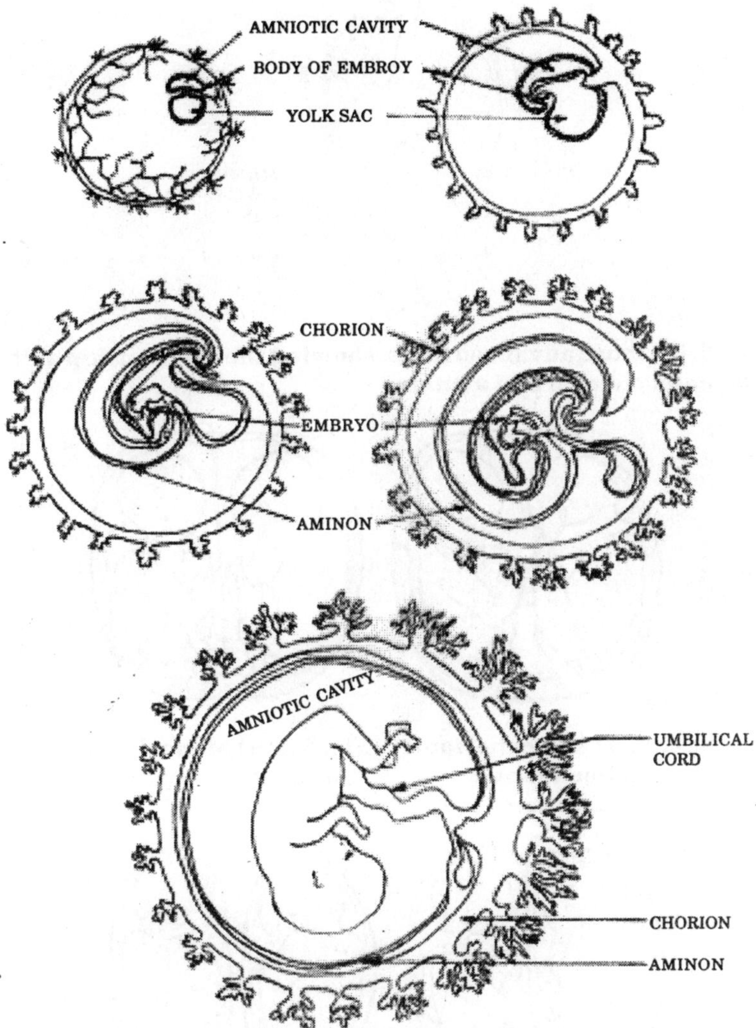

Fig. 2.3. The development of the embryo showing the derivation of the birth membranes. The amnion originates as the lining of the small amniotic cavity and eventually comes to surround the foetus. The chorion surrounds the amnion.

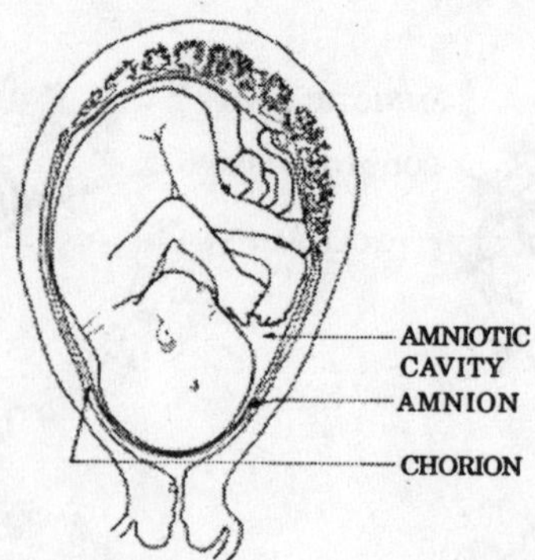

Fig. 2.4. A more advanced fetus, showing the relationship of the placenta, chorion and amnion.

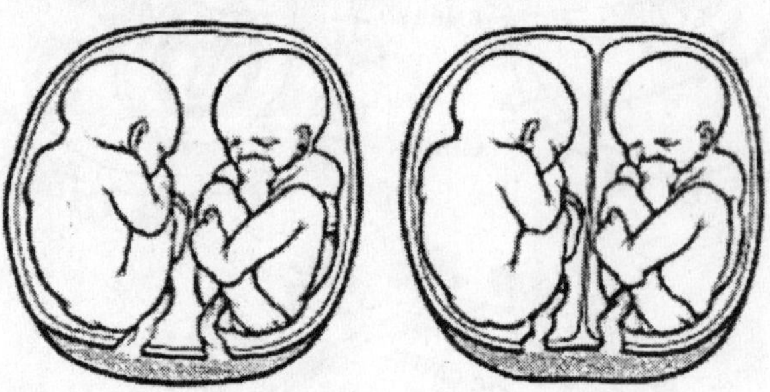

Fig. 2.5. (*i*) Monochorionic monoamniotic condition.

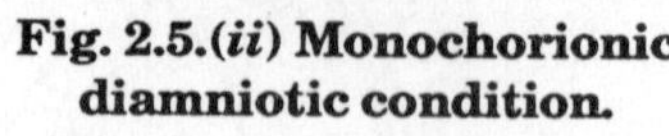

Fig. 2.5.(*ii*) Monochorionic diamniotic condition.

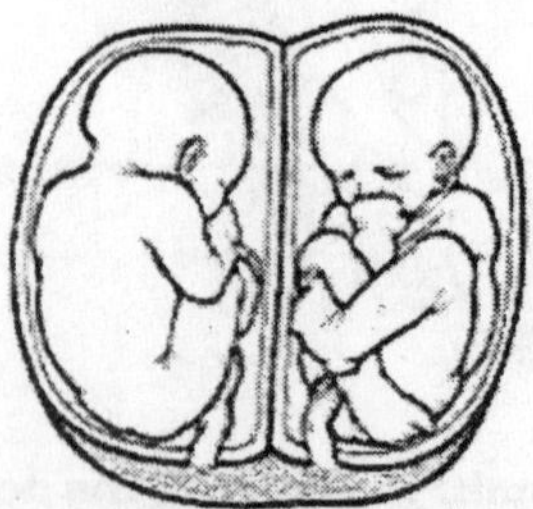

Fig. 2.5.(*iii*) Dichorionic diamniotic condition (fused placentae)

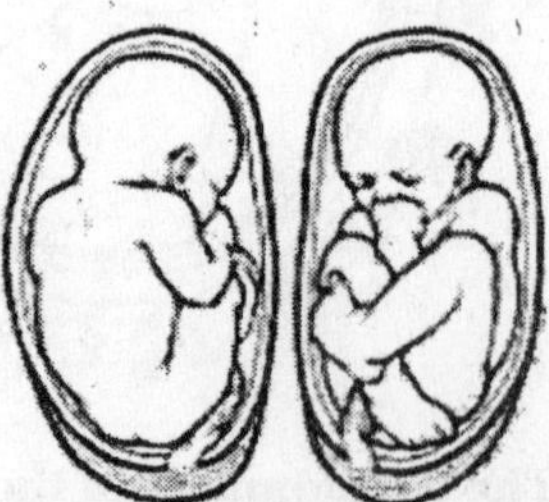

Fig. 2.5.(*iv*) Dichorionic diamniotic condition (separate placentae)

Fig. 2.5. Types of amnion and chorion in twin pregnancies.

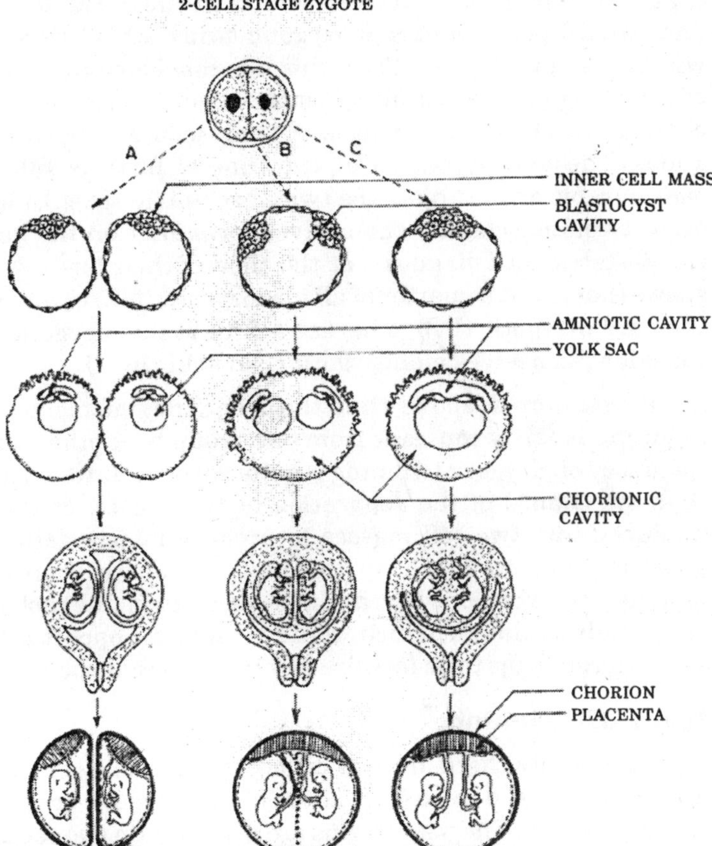

Fig. 2.6. Schematic diagrams showing the distribution of the birth membranes in monozygotic twinning. (A) The separation of the two cell masses is complete, so that each embryo has its own amnion,chorion and placenta.Sometimes the two placentas fuse to form a single one. This sequence is also characteristic of dizygotic twinning. (B)The two inner cell masses separate,but within a single blastocyst cavity. Each embryo then has its own amnion, but both are within a single chorion chorion.(C) The inner cell mass does not separate until later so that both embryos are included within a common amnion and chorion.

It is reported that out of 100 per cent of twins of an experiment of twins, 70 per cent of dizygotic twins and 10 per cent of monozygotic twins were found within a double chorion and rest 20 per cent of monozygotic twins, which were found within a single chorion. From this it is concluded that the type of twinning can be definitely established. If there is a single chorion, such type of twinning is certainly monozygotic (Bergsma, 1965). The determination at the time of birth is subject to considerable error (unless the twins are unlike-sexed) and very often twins misclassify themselves, primarily on the basis of the obstetrician's diagnosis at the time of their birth. A study shows that about a quarter of all monozygotic (MZ) twins believe they are dizygotic (DZ), whereas a fifth of the dizygotic twins consider themselves monozygotic (Novitski, 1977).

It was noted above that the frequency of monozygotic twinning is fairly constant from one group to another and the frequency of dizygotic twinning is quite variable. It would appear that the chance of the separation of the zygote, or its early products, into two cell masses to produce monozygotic twins does not depend much on the genetic constitution of the individuals involved, whereas, in dizygotic twin, the probability of multiple ovulation, which would present the opportunity for two different eggs to be fertilized by two different sperms, does.

Types of Twinning

Fundamentally there are two different kinds of twins such as (*a*) usual twin and (*b*) unusual twin of monozygotic or dizygotic origin (Fig. 2.7). The usual twins are divided into two categories like (*i*) monozygotic and (*ii*) dizygotic twins. But unusual twins of monozygotic or dizygotic origin are divided into five types such as (*i*) mosaicism, (*ii*) chimera, (*iii*) superfecundation, (*iv*) monozygotic twins of different phenotypes and (*v*) conjoined or Siamese twin. Fraternal twins are of no greater genetic relationship to each other than sibs born at different times; about half the time they are of the same sex. Identical twins, on the other hand, come from a single fertilized egg that has divided during the course of early development into two cells or two cell masses, each forming a new individual. Such twins are genetically identical to each other, with rare exceptions.

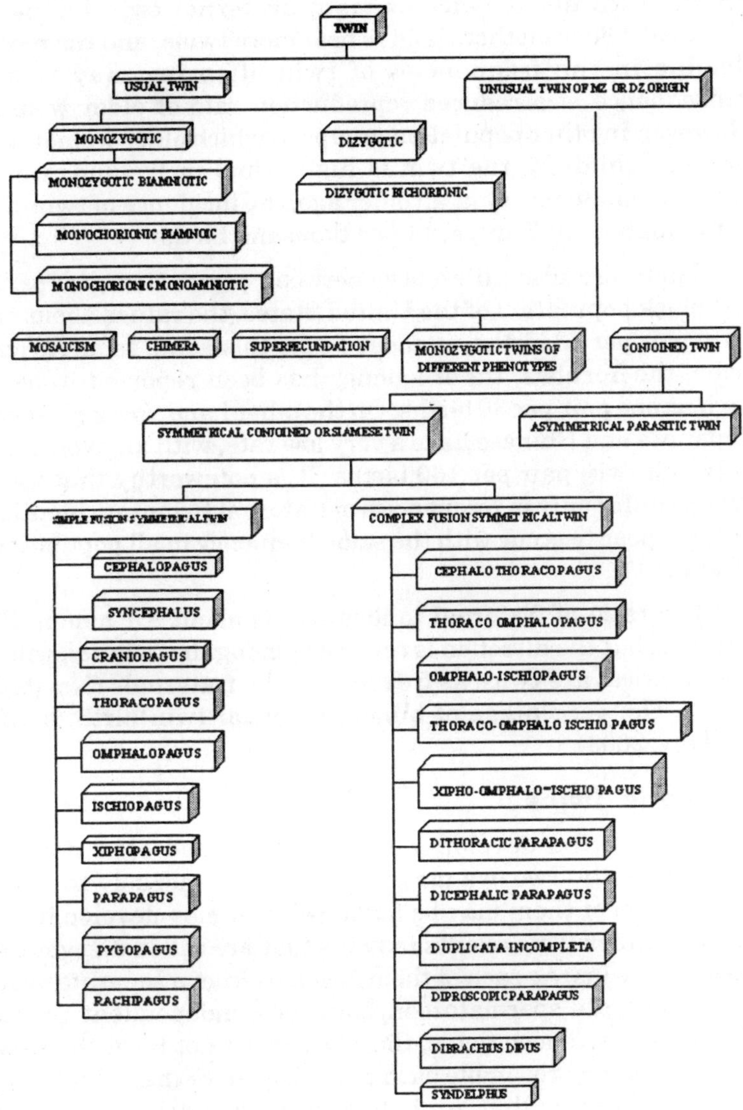

Fig. 2.7. Classificatio of human twins

The frequency of twinning in the white population of the United States is slightly more than one twin pair per hundred births, with about twice as many fraternal twin births as identical. Older mothers tend to bear more twins, and the recent decline in the frequencies of twin offspring may be the consequence of a reduced reproduction rate of older women. However, in other population groups in which older women still produce children, the twin is high. Thus in Ireland, where women tend to marry at an older age, the incidence of twinning is the highest in Europe, 14 per thousand births.

There are also differences between population groups. In the black population of the United States, the rate is about one twin pair per 73 births, or about 14 per thousand. In an African tribe, the Yorubas, the frequency has been reported to be as high as one pair per 30 births. On the other hand, among Asians, Japanese and Chinese have a very low rate, with an average on only one twin pair per 160 births. It is noteworthy that these groups differ only in frequencies of fraternal twinning. Identical pairs appear to occur with the same frequency in all populations (Table 2.1).

The ratio of fraternal to identical is about 8:1 among the Yorubas, but the direction is reversed among the Asiatics, where the identical are twice as frequent as the fraternals (Novitski, 1977). The percentage of human identical twin birth is 0.25 (Gilbert, 2003).

Dizygotic Twins

Dizygotic twins develop when two fertilized eggs develop independently. Two ova may be shed simultaneously from the ovary. Each of them may be fertilized, and may develop in the usual manner. This results in twins that are called dizygotic or fraternal twins. As each of them develops from a separate ovum and a separate spermatozoon, they have independent genetic constitutions. These twins, therefore, need not be of the same sex, nor do they resemble each other any more than children of the same parents that are born separately. Each fetus has its own chorionic and amniotic sacs. After the fertilized eggs are implanted into the endometrium, separate chorion and amnion develop for each egg, and later each conceptus develops its own placenta with an independent vascular system. If the fertilized eggs are implanted at a distance from each other, their placentas are absolutely separated (Fig. 2.5.iv and Fig. 2.8). If, however, the eggs are near each other, the contact between the placental margins is so tight that they seem to form a single placenta. This fusion is only apparent and each placenta is independent with its own vascular system (Fig. 2.9). This can be revealed by a close examination; each conceptus has its own amnion and chorion. The partition separating the two conceptuses consists of four membranes; two amnions and two chorions (Fig. 2.10). These membranes are easy to separate from each other. Dizygotic twins can be alike or unlike in sex.

Table. 2.1. Twinning rate per thousand births by race of mothers.

Sl.No.	*Name of the place*	*MZ*	*DZ*
1.	Indians of Arizona, New Mexico and Oklahoma	2.0	5.0
2	Oriental	4.5	2.5
3	U.S. White	3.8	7.4
4	U.S. Black	3.9	11.8
5	Johannesburg Black	4.9	22.3
6	Nigerian Black	5.0	39.9

Source: From tables by N.E. Morton, C.S. Chung, and M.P. Mi. Genetics of Interracial Crosses in Hawaii, S. Karger, AG, Basel, 1967; D. Hewitt and H. Stewart, *Acta. Genet. Med. Gemellol.*, 19 : 84, 1970.

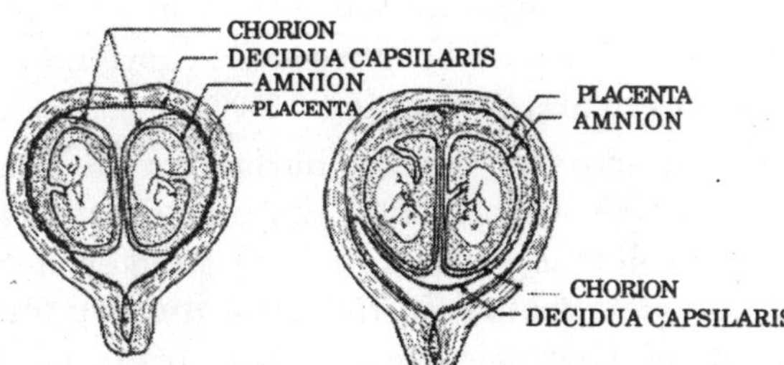

Fig. 2.8. Two eggs implanted in the uterus at a distance from one another : each foetus has its own placenta, amnion, chorion and decidua capsularis.

Fig. 2.9. Two eggs implanted near each other.; the placentas almost touch each other. Each foetus has its own amnion and chorion; the decidua capsularis is common.

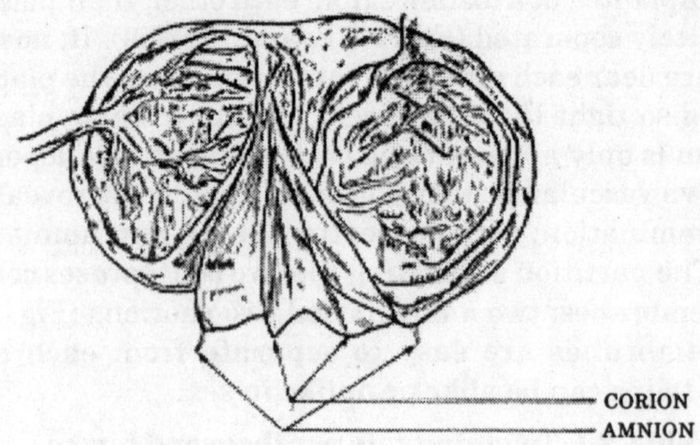

Fig. 2.10. The placenta of dizygotic twins. The partition separating the two fetuses comprises four membranes, viz., two amnions and two chorions.

Monozygotic Twins

Twins can also arise from a single fertilized ovum. These are called monozygotic or maternal twins. The genetic constitution of the two twins is exactly the same. Hence they are of the same sex and also exactly alike in appearance. Monozygotic twins are produced in one of the following ways.

(*a*) The cells formed in the first few divisions of the zygote are totipotent, *i.e.,* each cell is capable of developing into a complete embryo. The two cells formed by the first division may separate and develop independently. In such a case the fetuses bear separate chorionic and amniotic sacs, as in dizygotic twins.

(*b*) The embryo may develop normally upto the stage of the morula. However, when the blastocyst is formed, two inner cell masses form within it, and each develops into a complete foetus. In this case the two fetuses have a common chorionic sac but each lies in an independent amniotic cavity.

(c) Lastly, the inner cell mass may split into two or two embryonic axes may be established in one inner cell mass. By this it means that two separate embryonic discs are formed within it, each with its own prochordal plate and primitive streak. In such case the two fetuses share a common chorion as well as a common amniotic cavity.

In those instances where the fetuses share a common chorion, there is one placenta to which two umbilical cords are attached. Where the chorionic sacs are separate, two independent placentae are formed. Such placentae may secondarily fuse with each other, but normally there are no anastomoses between the vessels of the two placentae. Rarely, the placentae can fuse and there may be mixing of blood of the two fetuses. In that case the blood of each fetus may contain two types of erythrocytes. The condition is known as erythrocyte mosaicism.

The causative factors of monozygotic twinning are fascinating and depend on the stage of development at which the zygote splits to form two separate embryos. Accordingly, the monozygotic twins are classified under the following types such as (A) monozygotic bichorionic, (B) monochorionic biamniotic and (C) monochorionic monoamniotic twins.

A. Monogygotic Bichorionic Type

At two-cell stage of cleavage division, the totipotent blastomere cells develop into two separate zygotes within the common zona pellucida. When the zona pellucida disappears, the two zygotes are implanted separately in the uterine edometrium. Therefore, this variety of twins possesses separate placenta and separate chorionic sacs, like dizygotic twins (Fig. 2.5.(*iv*), Fig. 2.6A). About 25 per cent to 30 per cent identical twins are monozygotic bichorionic.

B. Monochorionic Biamniotic Type

At the blastocyst stage, the inner cell mass is separated completely into two equal parts and each part develops into a separate embryo. This variety of twins presents single placenta, common chorionic sac, but each embryo is invested by a separate amniotic membrane (Fig. 2.5.(*ii*) and Fig. 2.6.B, Fig. 2.11 and Fig. 2.12). About 70 per cent to 75 per cent identical twins are monochorionic biamniotic.

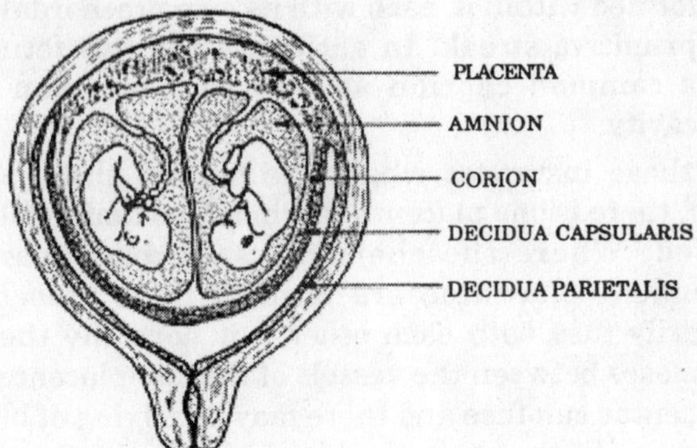

Fig. 2.11. Monozygotic(monoovular) twins. The placenta is common, each foetus has its own amnion.

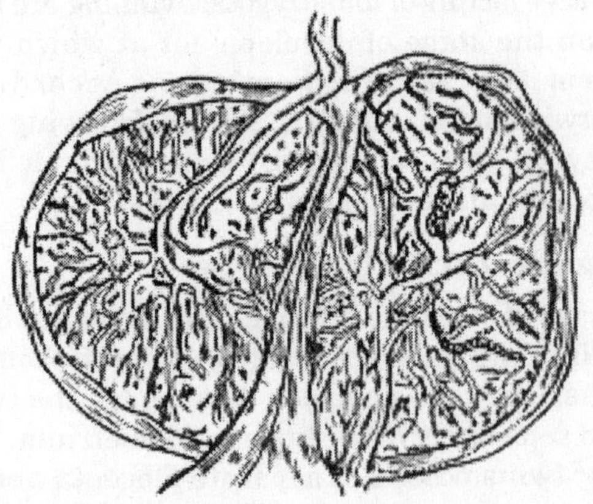

Fig. 2.12. Placenta of monozygotic twins. The partition separating the two fetuses consists of two membranes(two amnions).

C. Monochorionic Monoamniotic Type

When the bilaminar germ disc is formed within the blastocyst cavity, sometimes two separate organizing centers appear in the disc instead of one, and differentiate into separate embryos. The twin embryos lie within a common chorionic sac and are enveloped by a single amniotic membrane (Fig. 2.5.(*i*) and Fig.2.6.c). About one per cent to two per cent identical twins are monochorionic monoamniotic. Two possible hazards may be encountered in the monochorionic monoamniotic twinning namely (*i*) the conjoined twins or double monsters and (*ii*) the parasitic twins (Datta, 2004). The zygosity is determined by examining the placenta (common placenta, two separate, or contacting placentae) and the membranes. The factor, which is decisive for the diagnosis, is the number of leaves (Fig. 2.10 and Fig. 2.12) in the partition separating the conceptuses (four or two) (Bodyazhina, 1987).

Unusual Twin Events

The twins, which are not developed in usual manner, are considered as unusual twins. The unusual twins are divided into five types such as (*i*) mosaicism, (*ii*) chimera, (*iii*) superfecundation (*iv*) monozygotic twins of different phenotypes and (*v*) Siamese or conjoined twin.

Mosaicism

Mosaicism (Fr. *mosaique*, L.L. *masaicum*, Gr. *mousa*-a muse) is defined as the phenomenon of the presence of patches of tissues of unlike constitution side by side. A mosaic is simply a hybrid with the parental characters side by side, which is not blended. A mosaic consists of two different types of cells, originally the same at the stage of the zygote, but becoming genetically different subsequently by some event such as mutation, by nondisjunction in one cell line, or by X-chromosome inactivation. For instance, there are persons some of whose cells appear to be XX and other XY. Possibly there exist two twin zygotes initially of either sex. After one twin acquires cells from the other, the second twin embryo might have failed to become viable foetus. If the two cell masses fuse to give a single embryo, such a single birth shows two different kinds of cells would simply be defined as a mosaic.

The origin of mosaic embryonic mass of cells is by chromosome loss during first cleavage of XY zygote. This leads to two cell lines, one XY and the other XO (loss of Y chromosome). So that some of the resulting cells of XY embryo are XO (Fig. 2.13). On separation of this clump of cells into two groups, one of constitution XY and the other XO, two monozygotic twins of opposite sexes may develop. Monozygotic twins are sometimes mosaics. One of the unlike monozygotic pairs may consist of a male with short stature whose lymphocytes may be XO and a Turner's syndrome female may be a lymphocyte mosaic for XO and XY cells. It can be surmised that the male is also a mosaic of XO and ZY composition. This indicates the nondisjunction of chromosome Y in first cleavage of zygote. However, cleavage of cell mass through a plane leads to the development of two-cell mass that would make both the products mosaics (Fig. 2.14). Clearly since the cleavage might be in any direction, producing masses with unequal numbers of cells, the extent of mosaicism, if any, in each twin can be extremely variable.

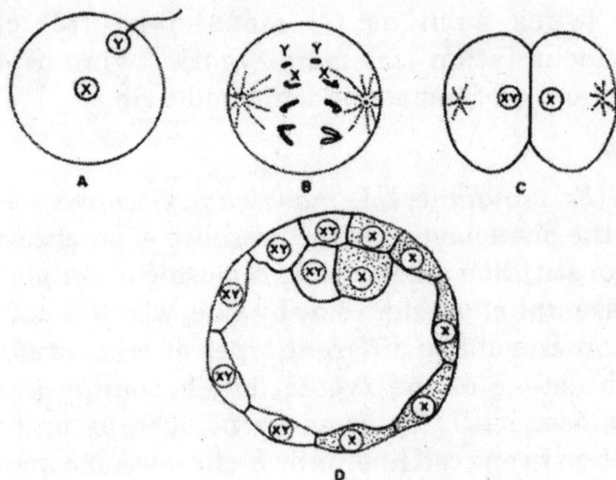

Fig. 2.13. Origin of a mosaic embryonic mass of cells after chromosome loss during the first cleavage of zygote.

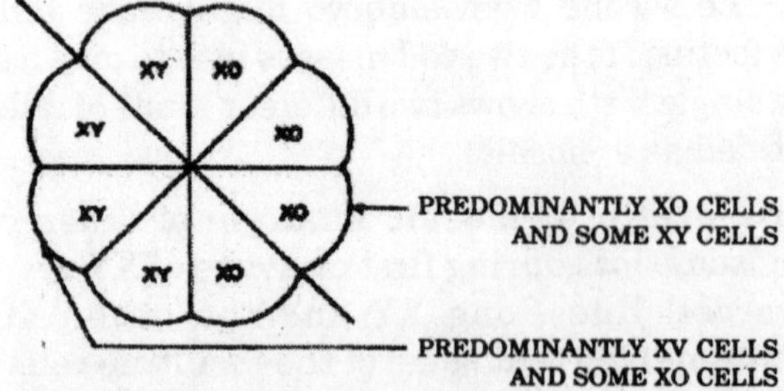

Fig. 2.14. The possible directions of cleavage and production of masses with unequal numbers of cells leading to mosaicism in each member of a twin.

Chimera

Chimera or chimaera (Gr. *chmaira-* a she-goat) literally means wild fancy or a picture of an animal having its parts made up of various animals. The general biological explanation of chimera is an organism, which is made up of two genetically distinct tissues. A chimera may be defined as an organism whose cells originate from two or more zygotes. The definition of blood chimerism requires the presence of cells derived from two (or more) genetically individuals populating the lympho-haemopoietic compartments. Blood chimerism may occur artificially or spontaneously.Artificial chimerism occurs transiently after blood transfusions. The spontaneous type of blood chimerism has been observed in twins after transfusion of haemopoietic stem cells between fetuses during first trimester.

However, in human dizygotic twins, cell exchange with proliferation in the genetically different foetus is rare, and only a few cases have been found. It is assumed that this blood chimerism is the result of an intensive reciprocal transfer of haemopoietic stem cells between the members of the twins and among the members of the triplets. Some transfused cells settle down in bone marrow of the host and produce blood cells of the donor line, a phenomenon that is well known from animal work and first described and supposed in the human. This interchange of blood cells occurred rather late in development by way of a fusion of blood vessels. This event is well known in cattle, in which dizygotic twins interchange cells in 90 per cent of the cases. When the sexes are different, the hormones from the male cause the female to develop into a sterile, somewhat male like type called a freemartin. Such an effect of one sex on the other in developing twins is unknown in humans.

Superfecundation

The phenomenon of production of members of same multiple births by different fathers is known as superfecundation and is well known in domestic animals. This is not reported in human beings.

Monozygotic Twins of Different Phenotypes

It is expected that if one member of an identical twin pair carries some unusual genetic trait, the other also shows as well. Identical twins having Down's syndrome, or both Klinefelter's syndrome, or both Turner's syndrome as well as three instances of twins having both Down's and Klinefelter's syndrome simultaneously have been reported (Novitski, 1977). In rarer instance, twins who satisfy all the criteria for monozygosity are strikingly different. Sometimes one appears to be male, the other female. Origin of these cases implies that mitotic nondisjunction may occur in XY zygote during early development (Fig. 2.15), giving rise to two cell lines, one XY and other XO (having lost

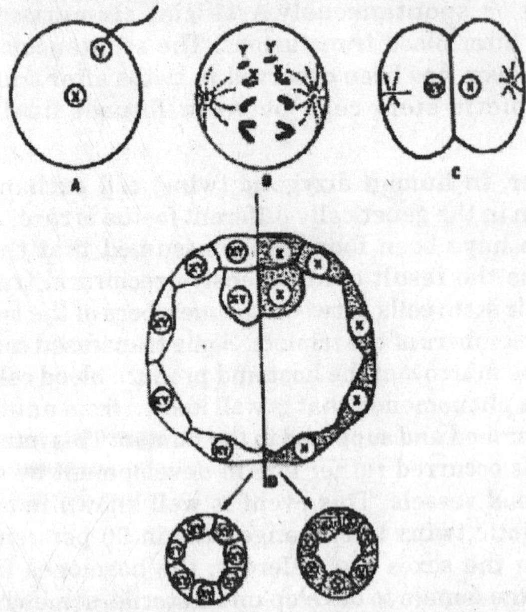

Fig. 2.15 Mitotic nondisjunction in first cleavage of zygote leading to two cell lines and separates to form one normal male; another female with Turner's syndrome of monozygotic origin.

the Y). Later separation of cell lines into two different masses occurs in such a way as to produce two embryos, one normal male and the other a female with Turner's syndrome, of monozygotic origin. In other cases monozygotic twins are like sexed but one is normal and other chromosomally abnormal. In this case if the zygotes are originally XX, then the loss of X chromosome could give rise to a twin pair, one normal female and the other having Turner's syndrome which is actually the most common type of chromosomally unlike monozygotic pair (Fig. 2.16).

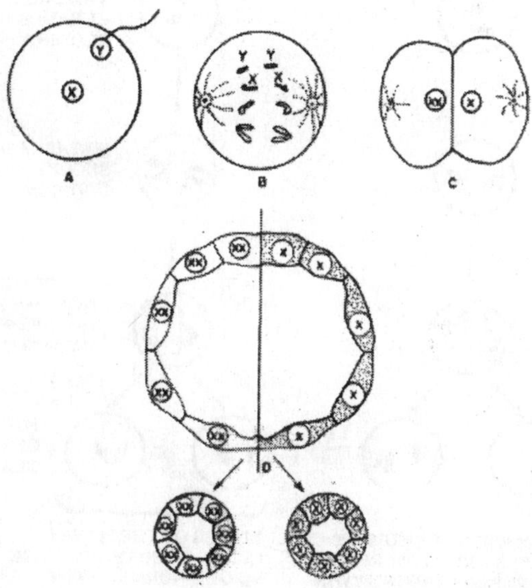

Fig. 2.16. Mitotic nondisjunction in first cleavage of zygote leading to two cell lines and separates to form one normal female and another female with Turner's syndrome of monozygotic origin.

In very few cases, it is seen that monozygotic twin pairs have been found, in which one is normal and the other with Down's syndrome. The course of events in this case is described as follows. After one or more normal mitoses of the zygote, a non-disjunctional event in a dividing cell gives rise to two different cell lines, one with three chromosomes 21 and the other with one. In fact, three different cell lines comes to picture. The monosomy 21-cell type reproduces slowly or not at all and eventually lost. The splitting of the cell masses made up of normal cells and trisomy 21 cells then occurs, creating twins along lines that results in one of them having primarily disomy 21 and the other trisomy 21 (Fig. 2.17).

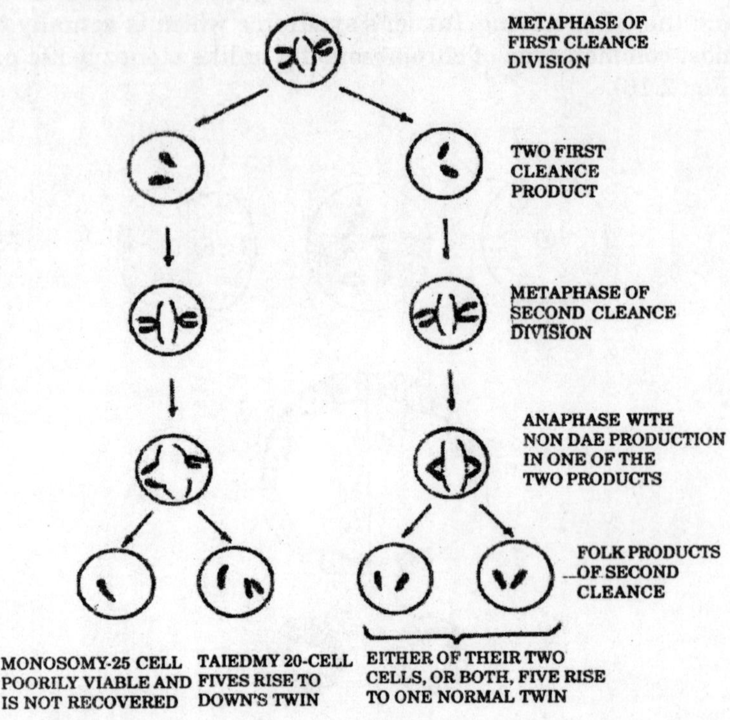

Fig. 2.17. A series of early cleavage divisions with nondisjunction of chromosome-21, giving rise to two monozygotic twins, one with trisomy-21 and the other normal.

Siamese or Conjoined Twins

Conjoining is classed as a congenital (development) defect. However, genes are involved in controlling development and development goes wrong due to malfunction of genes. Drugs and chemicals that seem to confuse the embryo, perhaps by mimicking substances that control embryo development, can also disrupt development. Two possible hazards may be encountered in the monochorionic monoamniotic twinnings, the conjoined twins or double monsters, and the parasitic twins.Conjoined twins are identical twins who develop with a single placenta from a single fertilized ovum and are always of the same sex.

Conjoined or Siamese twins arise by incomplete separation of fertilized ovum at about 15 days or more after zygote formation. Such twins are almost equal in size and receive proper blood supply. In this Siamese twins, the two partners may be completely separated by the appropriate surgery. Most of the conjoined twins do not survive due to complication at parturition.

According to the site and symmetry, the conjoined twins may be classified as symmetrical Siamese twin and asymmetrical parasitic twin, where one partner of the conjoined twins may be rudimentary due to diminished blood supply, and grows like a parasite from the body of the well-developed co-twin. Basing on degree of fusion, symmetrical conjoined twins are again divided into simple fusion Siamese twins and complex fusion Siamese twins (Table 2.2). Simple fusion Siamese twins may be classified into 10 categories such as :

(1) `cephalopagus,
(2) syncephalus,
(3) craniopagus,
(4) thoracopagus,
(5) omphalopagus,
(6) ischiopagus,
(7) xiphopagus,
(8) parapagus,
(9) pygopagus and
(10) rachipagus. Complexion fusion of Siamese twin may be grouped into 11 categories such as :
(1) cephalothoracopagus,
(2) thoraco-omphalopagus,
(3) omphalo-ischiopagus,
(4) thoraco-omphalo-ischiopagus,
(5) xipho-omphalo-ischiopagus,
(6) dithoracic parapagus,
(7 dicephalic parapagus,
(8) diproscopic parapagus,
(9) duplicata incompleta
(10) dibrachius dipus and
(11) syndelphus.

Table 2.2. Types of conjoined twins and their morphological features

Sl. No.	*Type*	*Site of fusion portion of the body*	*Morpholigical and anatomical feature (s)*
1.	Cephalopagus	Anterior union of the upper half of the body and heart may be conjoined or shared.	Two faces on opposite sides, 4 arms and 4 legs.
2.	Syncephalus	Joined by the face	It has 1 head, 1 face 2 bodies and 4 ears, out of which 2 ears on the back of the head.
3.	Craniopagus	Joined at the skull but not at the face or base of the skull	Shared bones of the skull and often have fused brain surffaces, 4 arms and 4 legs.
4.	Thoracopagus	Joined face to face from the upper chest to the navel with a shared or conjoined heart.	2 heads, 4 arms, 4 legs and 2 pelvises.
5.	Omphalopagus	Joined face to face at the navel and some times also lower chest and liver may be conjoined.	2 heads, 2 separate hearts, 2 pelvises, 4 arms and 4 legs.
6.	Ischiopagus	Joined from the navel to a large conjoined pelvis joined end to end with the spine in a straight line.	4 heads, 4 arms, variable number of legs, usually single external genitalia and single anus.
7.	Xiphopagus	Joined at the xiphoid process and shared or conjoined liver.	2 heads, 4 arms, 4 legs and separate vital organs.
8.	Parapagus	Conjoined pelvis fused side by side.	2 heads, 2 chests, arms variable (2 or 3 or 4) and legs also variable (2 or 3 or 4).
9.	Pygopagus	Joined at the perineum and many share part of the spinal cord.	2 heads, 4 arms, 2 rectums and a single anus.

Sl. No.	*Type*	*Site of fusion portion of the body*	*Morpholigical and anatomical feature (s)*
10.	**Rachipagus**	**Twins fused back to back above the sacrum involving different segments of spines.**	**2 heads, 4 arms and 4 legs.**
11.	**Cephalothoraco-pagus**	**Fused from head to chest, share heart and fused gastro-intestinal tract.**	**1 brain, 1 head, there may be rudimentary 2nd face, 4 arms and 4 legs.**
12.	**Thoraco omphalo-pagus**	**Conjoined from upper chest to the navel, shared a single heart or have conjoined heart and have shared liver.**	**2 heads, 4 arms and 4 legs.**
13.	**Omphalo-Ischi-pagus**	**Face to face with a joined abdomen and fused pelvis.**	**2 heads, 4 arms and variable number of legs.**
14.	**Thoraco-omphalo-Ischiopagus**	**Joined chest, joined abdomen and fused pelvis.**	**2 heads, 4 arms and variable number of legs.**
15.	**Xipho-omphalo-Ischiopagus**	**Face to face with a joined lower chest, abdomen and fused pelvis.**	**2 heads, 4 arms and variable number of legs.**
16.	**Dithoracic para-pagus**	**Joined at abdomen and pelvis.**	**2 heads, 2 chests and 4 arms and variable number of legs (2 or 3 or 4)**
17.	**Dicephalic para-pagus**	**Joined at chest, abdomen and pelvis**	**One trunk, 2 heads, 4 arms and 2 legs.**
18.	**Diproscopic para-pagus**	**Joined at chest, abdomen and pelvis**	**1 head with 2 faces.**
19.	**Duplicata incom-pleta**	**Any part which has not completely separated.**	**2 heads and 1 body.**
20.	**Dibrachius dipus**	**Fused head at any portion of the cephalic region.**	**2 arms, 2 legs *i.e.*, "normal body layout.**
21.	**Syndelphus**	**Fused head but single body.**	**1 body and 8 limbs**

Simple Fusion Siamese Twin

Cephalopagus (Gr. *kephale*-head; *pagus*-combining form)

In cephalopagus twin, the anterior (front of the body) upper half of the body gets fused with two faces on opposite sides of a conjoined head. The heart may also be conjoined or shared (Fig.II.18.i). Their survival is very difficult.

Syncephalus (Gr. *syn*-together; *kephale*-head; *pagus*-combining form)

A form of cephalopagus twins; joined by the face and has a single head and two bodies are described as syncephalus twin. Syncephalus twins usually have one head and a single face with four ears, two on the back of the head and the other two in the normal position.

Craniopagus (L.L. *cranium*, Gr. *kranion* – the skull; *pagus*-combining form)

Craniopagus twins are joined at the skull, but not at the face or base of the skull, and share bones of the skull and often have fused brain surfaces (Fig. 2.18.(*ii*)). They can be joined crown to crown (vertical), at the back of the head (occipital), front of the head (frontal) or the side of the head (parietal or temporal).

Thoracopagus (Gr. *thorax*-chest; *pagus*-combining form)

Joined face to face from the upper chest down to the navel, with a shared heart of conjoined heart, four arms, four legs and two pelvises are the characteristic features of thoracopagus twins (Fig. 2.18.(*iii*)).

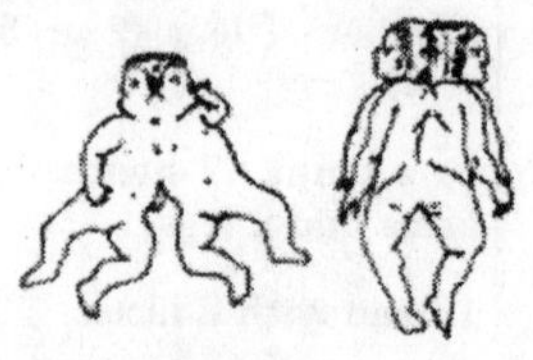

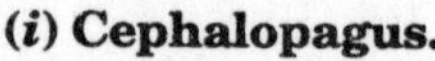

(*i*) Cephalopagus.

(*ii*) Craniopagus.

(*iii*) Thoracopagus.

(*iv*) Omphalopagus.

(*v*) Ischiopagus.

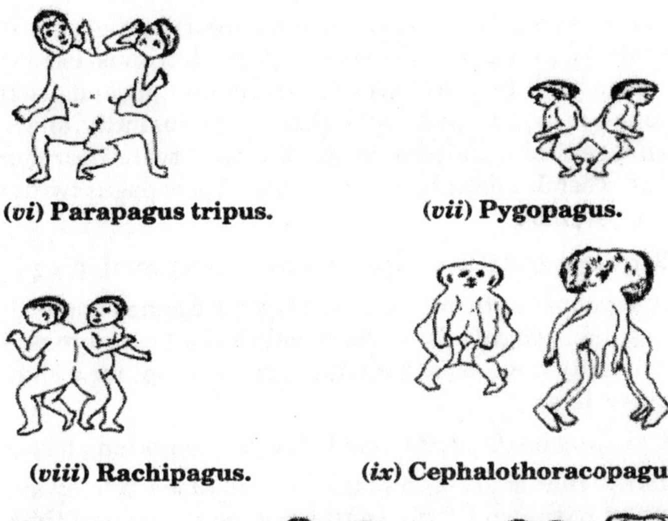

(*vi*) Parapagus tripus. **(*vii*) Pygopagus.**

(*viii*) Rachipagus. **(*ix*) Cephalothoracopagus.**

(*x*) Thoraco-Omhalopagus

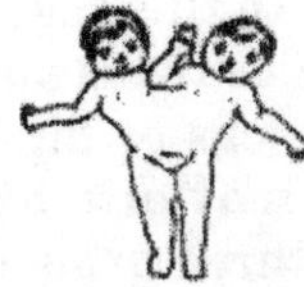

(*xi*) Dithoracic tetrabrachius dipus.

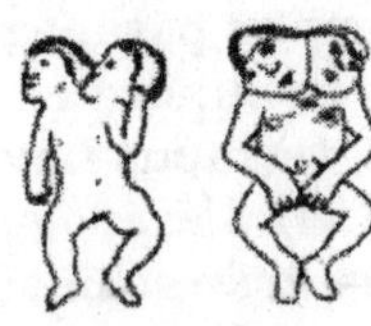

(*xii*) Dicephalus dibrachius dipus.

Fig. 2.18. Types of symmetrical Siamese or conjoint twins.

Omphalopagus (Gr.*omphalos*-navel; *pagus*-combining form)

The unique features of omphalopagus twins are as follows. They are joined face to face at the navel and sometimes also the lower chest. Other structural features are two separate hearts, but the pelvis may be conjoined, two pelvises, four arms and four legs (Fig. 2.18.iv).

Ischiopagus (Gr. *ischion* – the hip joint; *pagus*-combining form)

Joined from the navel to a large conjoined pelvis joined end to end with the spine in a straight line are the features of ischiopagus twin. They have four arms, a variable number of legs, usually single external genitalia and a single anus (Fig.

2.18.v). The variations of ischiopagus are (*i*) ischiopagus dipus, (*ii*) ischiopagus tripus and (*iii*) ischiopagus tetrapus. Ischiopagus dipus has only two legs and can resemble parapagus twins, ischiopagus tripus possesses three legs, usually one set of external genitalia and ischiopagus tetrapus twins bear four legs and can resemble dicephalus or dithoracic parapagus twins when it sits or stands.

Xiphopagus (Gr. *xiphos*-a sword; *pagus*-combining form)

In xiphopagus twins, the individuals are joined at the xiphoid process, i.e., roughly from the navel to the lower breast bone. They usually have separate vital organs except for a shared or conjoined liver.

Parapagus (Gr. *para*-beside ; *pagus*-combining form)

Twins that share a conjoined pelvis fused side by side are known as parapagus twin. In this case some parts of the pelvis may be duplicated. Parapagus twins may be tripus having three legs, tetrapus having four legs or dipus having two legs (Fig. 2.18.vi). Parapagus twins may also be dibrachius having two arms, tribrachius having three arms or tetrabrachius having four arms.

Pygopagus (Gr. *pyge*-rump; *pagus*-combining form)

Pygopagus twins are identified from the joining at the perineum and may share part of the spinal cord. They have a single anus, two rectums, four arms and legs (Fig. 2.18.vii).

Rachipagus (Gr. *rachis* – spine; *pagus* combining form)

Twins fused back to back above the sacrum, involving different segments of the spine are categorised as rachipagus twin (Fig.III.18.viii).

Complexion Fusion of Siamese Twins

Cephalothoracopagus (Gr. *kephale*-head; *thorax*-chest; *pagus*-combining form)

Fused from the head to chest is the cephalothoracopagus twin. There is only one brain and a head. There may be a rudimentary second face. They usually share a heart and have fused gastro-intestinal tracts and are non-viable. They are similar to syncephalus, but the chest is fused (Fig. 2.18.ix).

Thoraco-Omphalopagus (Gr. *thorax* – chest; *omphalos*–navel; *pagus*-combining form)

Thoraco-omphalopagus twin is a combination of thoracopagus and omphalopagus. They are conjoined from the upper chest to the navel and either share a single heart or have conjoined hearts. They also have shared livers and gastro-intestinal tracts, four arms and four legs (Fig. 2.18.x).

Omphalo-Ischiopagus (Gr. *omphalos*-navel; *ischion*- the hip joint; *pagus*-combining form)

Omphalo–ischiopagus is the combination of omphalopagus and ischiopagus. They are joined as ischiopagus twins, but face to face with joined abdomen and fused pelvis. They have four arms and a variable numbers of legs (2 or 3 or 4).

Thoraco-Omphalo-Ischiopagus (Gr.*thorax*-chest; *omphalos* navel; *ischion* – the hip joint; pagus-combining form)

Thoraco–omphalo–Ischiopagus are just like omphalo-ischiopagus but also joined at the chest. They bear two heads, four arms and a variable numbers of legs (two or three or four).

Xipho-Omphalo-Ischiopagus (Gr. *xiphos* – a sword; *omphalos* – navel; *ischion* – the hip joint; *pagus*-combining form)

Xipho-omphalo–ischiopagus are similar to omphalo–ischiopagus but they are joined from the lower chest. They have two heads, four arms and a variable number of legs (two or three or four).

Dithoracic Parapagus (Gr. *dis* – twice; *thorax* – chest; *para* – beside; *pagus*-combining form)

Dithoracic parapagus are joined at the abdomen and pelvis, but not the chest. They are characterised by the presence of two chests, two heads, four arms and two legs (Fig.II.18.xi).

Dicephalic Parapagus (Gr. *dis* – twice, *kephale* – head; *para* – beside; *pagus*-combining form)

Dicephalic parapagus has one trunk with two heads. They are also characterised by presence of two arms or four arms and two legs (Fig. III. 18.xii).

Diproscopic Parapagus (Gr. dis – twice; *pro* –before; *skopeein*- to see; *para* – beside; *pagus*-combining form)

Diproscopic parapagus has a single head with two faces.

Duplicata Incompleta (L. *duo*, two; *plicare* – to fold; *in*- not; *completus* – filled, finished)

Duplicata incompleta twins are observed as any part, which has not completely separated. They are dicephalus having two heads and one body.

Dibrachius Dipus: (Gr. *dis*– twice ; *brachion* – arm; *dis* – twice; *pous, podos*-foot)

Dibrachius dipus twins are characterised by two arms, two legs, *i.e.,* "normal" body lay out.

Syndelphus: (Gr. *syn* – together; *adelphos*-brother)Syndelphus is described as twins having one body and eight limbs.

Parasitic Twins

Symmetrical Siamese twins are those in which one member is fairly normal (the host or auto site) but the other member is a parasite, which is severely underdeveloped and dependent on its host for nutrition. Such parasites may often be removed by surgery. One partner of the conjoined twins is rudimentary due to diminished blood supply (Fig. 2.19). Sometimes the parasite twin may be completely enclosed within the body of the co-twin. This phenomenon is known as "foetus in foetus".

Fig. 2.19. Asymmetrical Siamese or conjoint twin (Parasitic twin).

3

Traits of Twins

Different Traits of Human Twins

In addition to comparison between identical and non-identical twins, further comparison can be made between the members of identical pairs reared in different environments. Since identical twins have the same genetic constitution,rearing each member of a pair separately may show which traits are more or less susceptible to environmental differences and thereby such environment help to reveal the extent of genetic influence on different characters.

In this work, besides,the study of pedigree to know the inheritance of twinning,traits like body and eye colour,shape of ear and hair,vision,blood group, tongue folding and rolling, handedness, hand writing, intelligence and pitch of the voice of the twins have been studied. More equal concordance and discordance ratios between the identical and non-identical groups would,in turn, signify less emphasis on heredity similarity in determination of the trait and more emphasis on environmental factors.

Pedigree

In this study, pedigree has been analysed to trace the twin births in the ancestors. The inheritance of human traits like polydactyly, syndactyly, idiolgy, intelligence, skin colour, hemophilia, albinism, colorblindness and other diseases have been studied by pedigree analysis. The pedigree chart of a family showing inheritance of albinism as a recessive trait is exemplified (Fig. 3.1). In this figure circles (O) represent females and squares (□) designate males. The parents are connected by horizontal line (——) and the vertical lines (1) lead to their offsprings. The offsprings are called sibs and are connected by a horizontal sibship line. Sibs are placed from left to right according to birth order, and are labelled with Arabic numerals. Each generation is indicated by Roman numeral. Thus, a designation such as 1-2 represents the female of the first set of parents in the initial generation.

A diamond (□) is used (II-2) is unknown individuals. If a pedigree traces only a single trait, the circles, squares and diamonds are shaded (I-1, III-3 and III-4) when the phenotype is expressed (Fig. 3.1). The heterozygous who fails to express a recessive trait, is symbolized by half of the shaded square or shaded circles (II-3 and II-4).

Twins are indicated by diagonal lines stemming from a vertical line connected to the sibship line. For monozygotic twins, the diagonal lines are linked by horizontal line (III-5 and 6), whereas dizygotic twins lack this connecting line (III-8 and 9). A number within one of the symbols (III-10 and 13) represents numerous sibs of known or unknown phenotypes. The individual whose phenotype draws the attention of a physician or geneticist is called the proband and is indicated by an arrow connected to the designation p (III-4) and this term applies to both males and females. In this figure the proband happens to be a male.

The pedigree analyses of twins (Figs. 3.2 to 3.20) reveal the presence of twins in both maternal and paternal generations of twins and triplets. Thus, it indicates the contribution not only of the female but also of the male for twinning and multiple birth tendency. From this it is apparent that some hereditary factors are responsible for twinning.

Body and Eye Colour, Shape of Ear and Hair, Vision, Blood Group, Tongue Folding and Rolling and Handedness

Body colour in man is governed by genes. The presence of melanin pigment in the skin determines the skin colour. The more is the pigment, the darker is the skin. Usually five types of skin colours are found in man such as negroes, dark, intermediate, light and white. In this investigation, the body colour such as white, intermediate and dark colour have been observed in twins and 67 per cent of non–identical twins exhibit concordance in their body colour whereas 100 per cent of concordance is seen in identical twins. However,in relation to eye colour, it is found to be 80 per cent in non-identical twins and that to 100 per cent of concordance in identical twins.

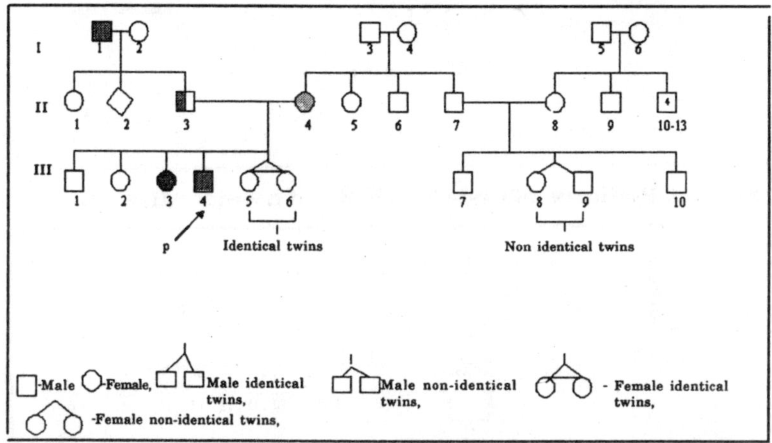

Fig. 3.1. Representative pedigree for a single character through three generations.

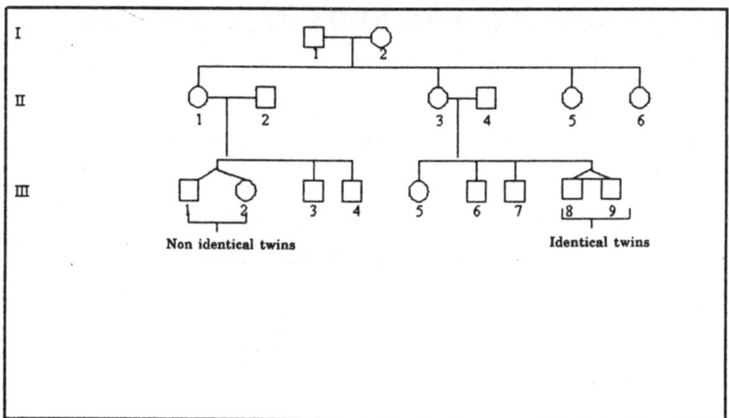

Fig. 3.2. Pedigree of Code No. GNJ 3 of non-identical twins.

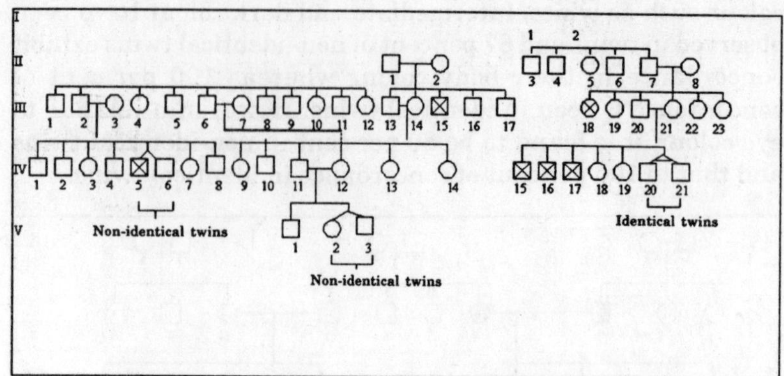

Fig. 3.3. Pedigree of Code No. JPR 1 of non-identical twins.

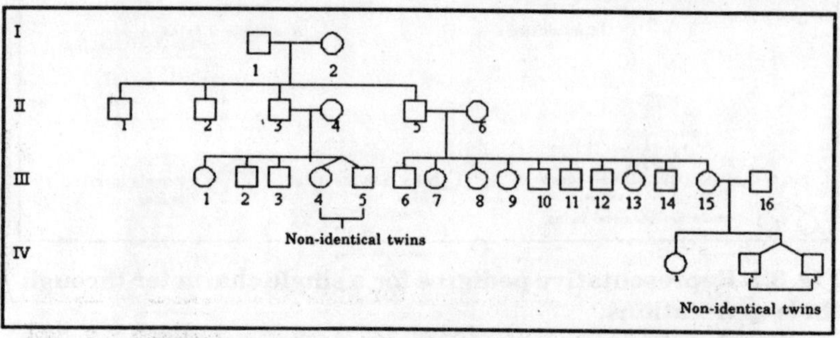

Fig. 3.4. Pedigree of Code No. KHU 3 of non-identical twins.

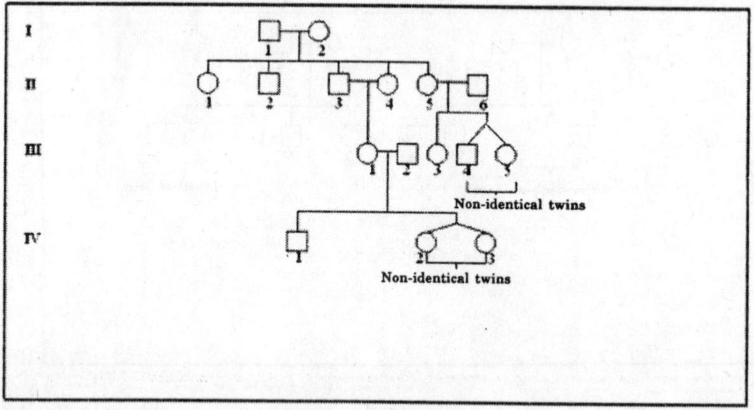

Fig. 3.5. Pedigree of code No. KHU 18 of non-identical twins.

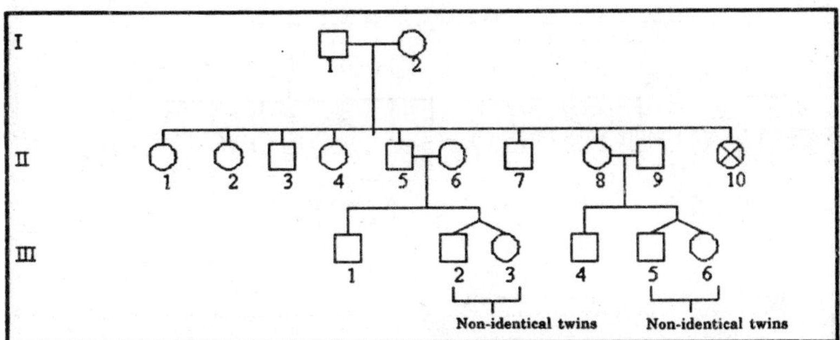

Fig. 3.6. Pedigree of Code No. KNP 2 of non-identical twins.

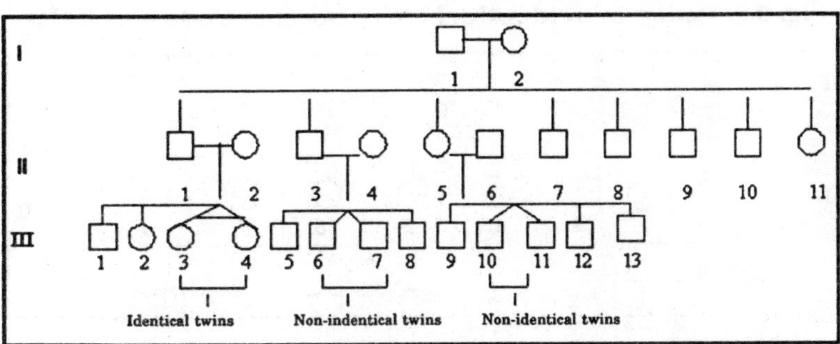

Fig. 3.7. Pedigree of code No. PUR 1 of non–identical twins.

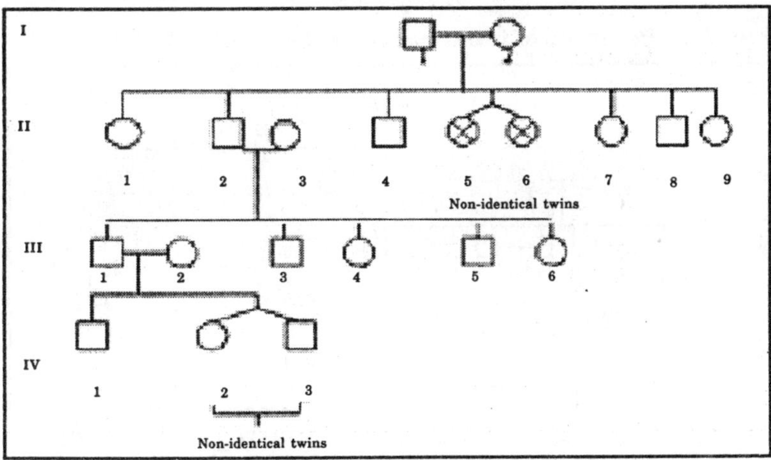

Fig. 3.8. Pedigree of code. PUR 3 of non-identical twins.

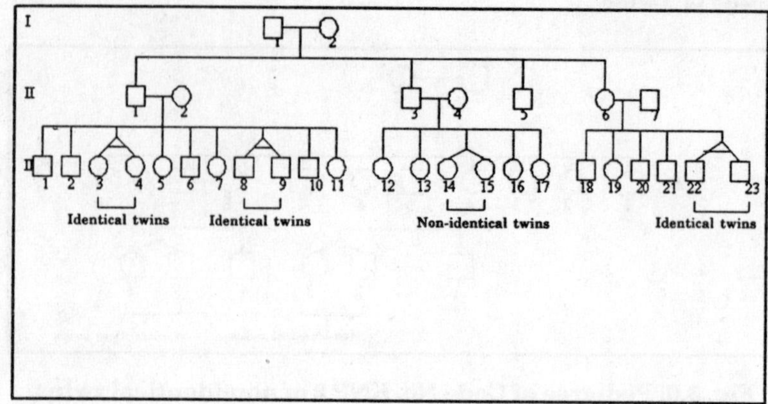

Fig. 3.9. Pedigree of identical twins bearing Code no. CTC 2.

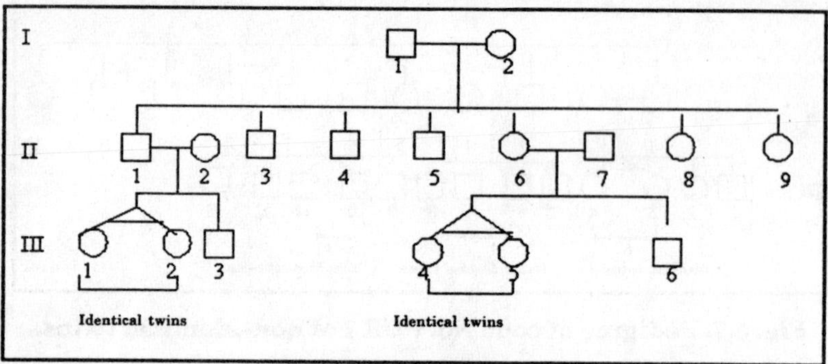

Fig. 3.10. Pedigree of identical twins bearing code No. GNJ 3.

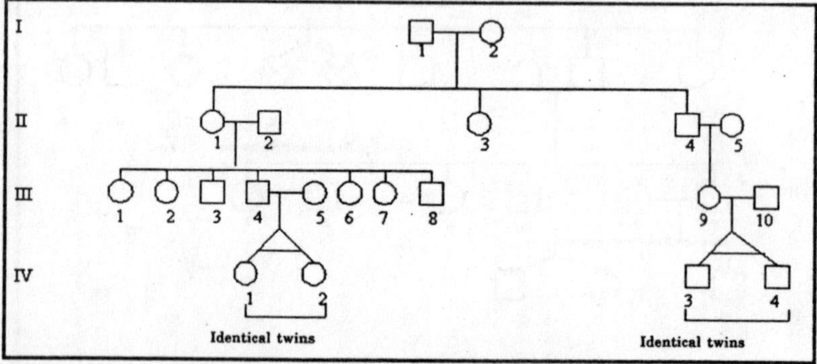

Fig. 3.11. Pedigree of identical twins bearing code No. JSP 1.

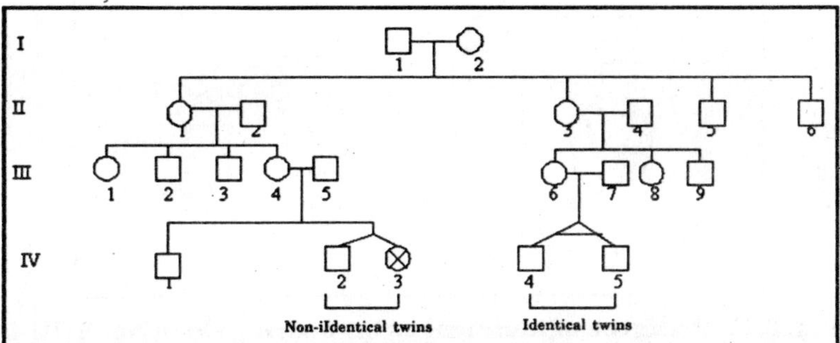

Fig. 3.12. Pedigree of identical twins bearing code No. JSP 2.

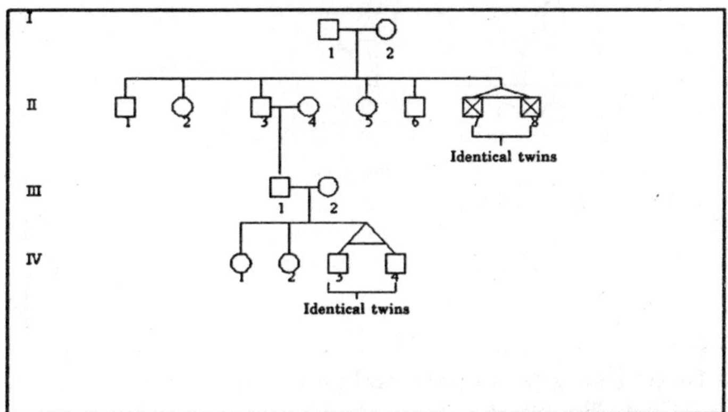

Fig. 3.13. Pedigree of identical twins bearing code No. JPR 1.

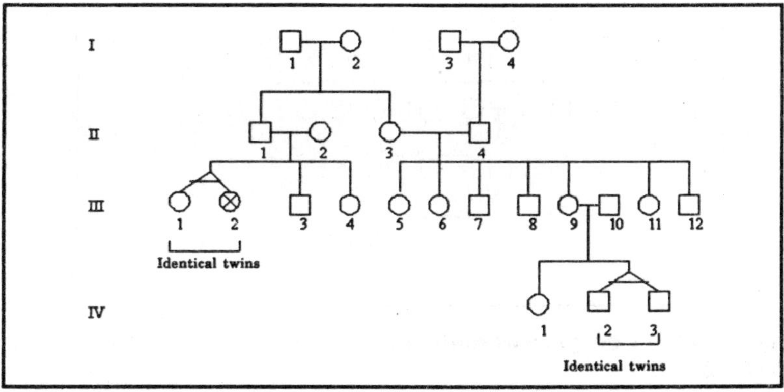

Fig. 3.14. Pedigree of identical twins bearing code No. KHU 3.

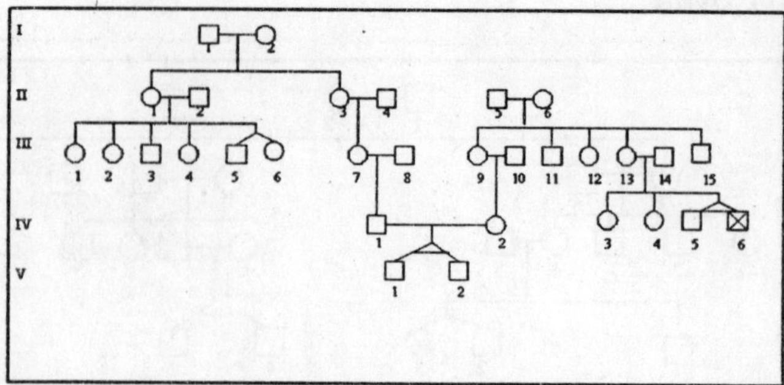

Fig. 3.15. Pedigree of identical twins bearing Code No. KHU 4.

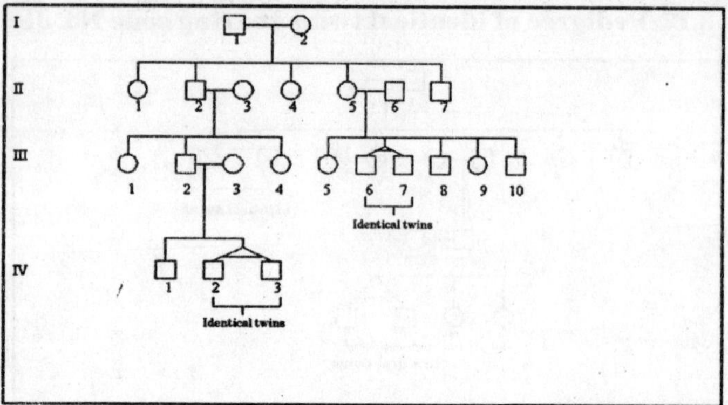

Fig. 3.16. (*i*) Pedigree of paternal generation of identical twins bearing Code No. NGH 2.

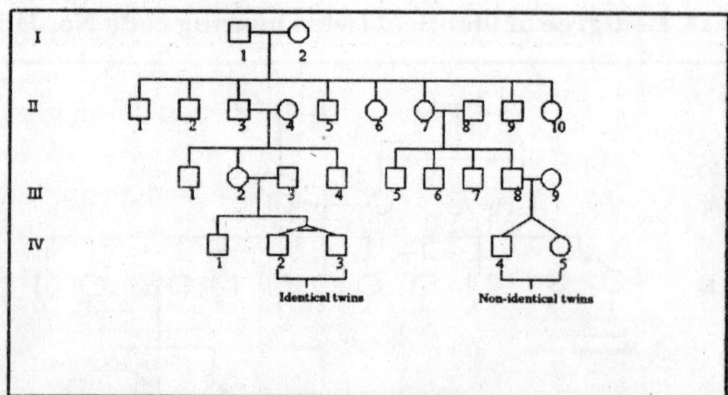

Fig. 3.16.(*ii*) Pedigree of maternal generation of identical twins (NGH 2).

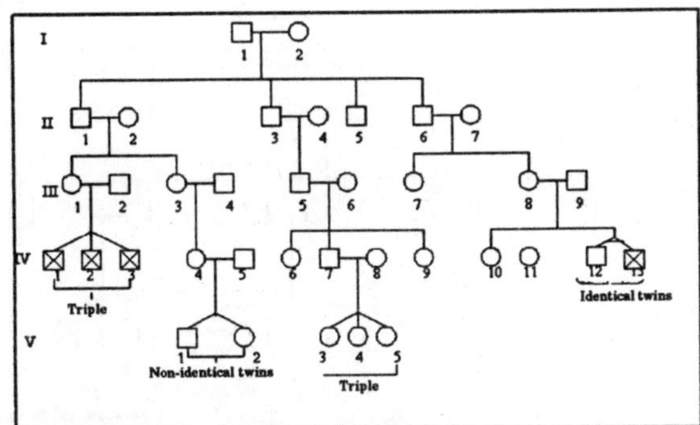

Fig. 3.17.(*i*) Pedigree of paternal generation of triplet (Code No. KHU 2).

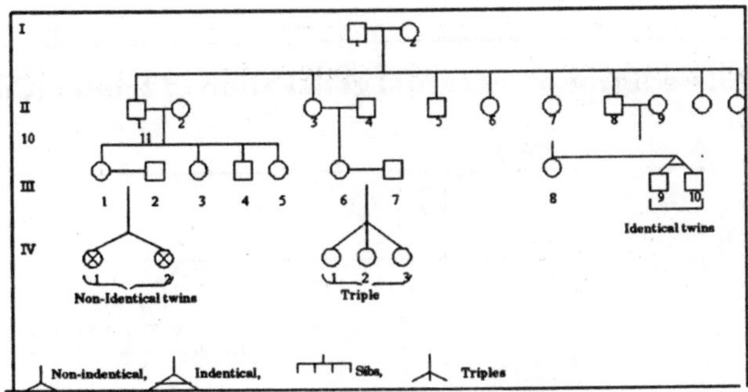

Fig.3.17.(*ii*) Pedigree of maternal generation of triplet (Code No. KHU 2).

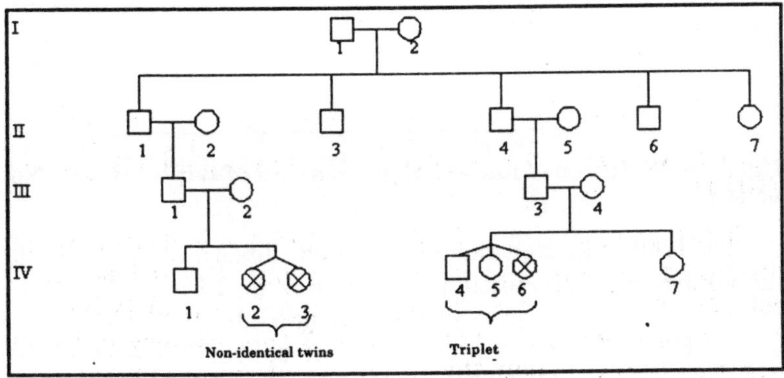

Fig. 3.18.(*i*) Pedigree of paternal generation of triplet (Code No. KHU 3).

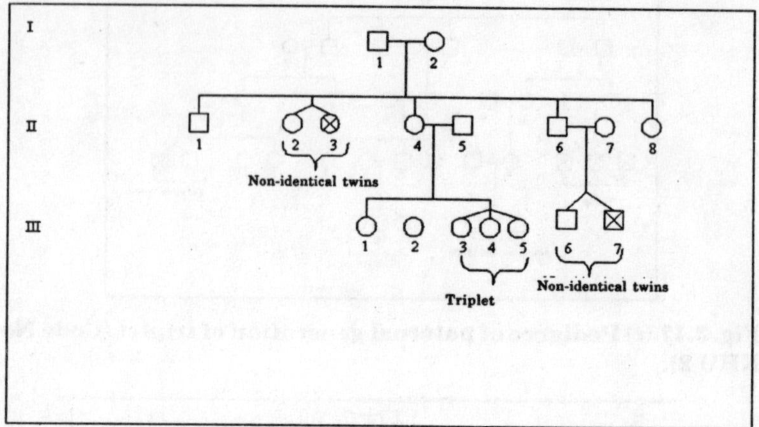

Fig. 3.19. Pedigree of maternal generation of triplet (Code No. KNP 1).

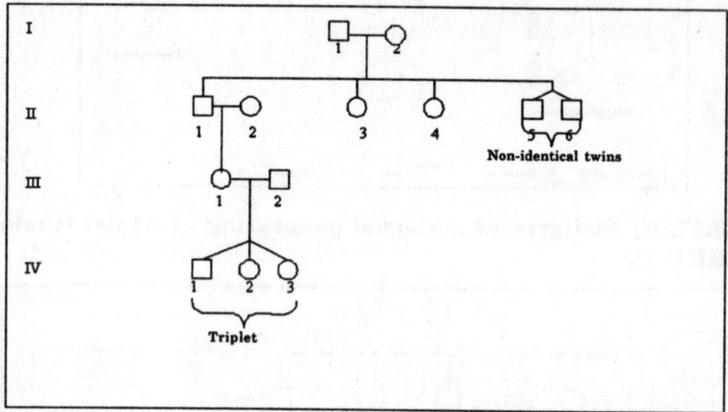

Fig. 3.20. Pedigree of maternal generation of triplet (Code. No. NGH 1).

The traits like free and adherent ear lobes and straight and curly hair are of hereditary nature in man. In ear lobe, 67.71 per cent of concordance is seen in non-identical twins and in identical twins the percentage of concordance is 90.38. Regarding shape of hair, the percentage of concordance is found to be 84.31 and 100 in non-identical and identical twins respectively.

The trait like vision shows normal and abnormal eye sight such as short and far sightedness in the members of the twins. However, 86.27 per cent of concordance is found in the vision of non-identical twins whereas identical twins exhibit 92.3 per cent.

Regarding blood groups of man, four blood groups have been differentiated basing on presence or absence of antigens and antibodies. These are designated as A, B, AB and O blood groups. The person belonging to blood group A have antigen-A in their red blood cells and antibody b (anti-b) in the plasma. Persons of blood group B have the antigen-B in their red blood cells and antibody a (anti-a) in the plasma. Persons of group AB have antigen A and B in the red blood corpuscles but no antibodies in the plasma. Persons of blood group O have no antigen in red blood cells but have both antibodies a and b in the plasma. Another antigen is designated Rh-positive (Rh+) and those without it are Rh-negative(Rh-).

In relation to blood group A+, B+, AB+, AB- and O+ blood groups have been observed in the members of twins and 29.41 per cent of non-identical twins show concordance in their blood groups and that to 100 per cent of concordance is seen in identical twins.

The traits like tongue folding, tongue rolling and handedness such as righty and lefty are behavioural traits in man. As to tongue folding 71.43 per cent of non-identical twin and 88 per cent of identical twins show concordance in such trait.

Study on tongue rolling shows 73.47 and 94 per cent of non–identical and identical twins are found to be concordant in tongue rolling respectively.

Regarding handedness, members of twins show righthanded and lefthandedness and 90.2 per cent of non-identical twins and 94.23 per cent of identical twins show concordance in such trait.

Handwriting of Twins

The handwriting of twins exhibits identifying features which is hereafter referred to as its "characteristics". The characteristics of a handwriting fall into two classes such as (i) those derived from the general style to which the handwriting conforms, are termed as "style characteristics", and (ii) those which have been introduced into the handwriting whether consciously or unconsciously, by the writer. These are referred to as " personal characteristics".

Handwriting is developed gradually over the ages. The variation in the relative sizes of letters and parts of letters plays such an important part in the comparison of handwriting. The value of this determination of the nature and extent of the variation around the master pattern becomes evident when two handwritings which have similar master patterns for any particular letter are considered.

In twins study (Figs. 3.21 to 3.24), handwriting of twin members are compared from the size, spacing, variation, style, initial stokes, terminal spurs, connecting strokes, slope, shading and relative alignment of letters.

In relation to handwriting identical twins show a resemblance particularly when they are reared together. In general, the pattern of handwriting of identical twins reared together differs from that of twins reared apart. From this, it is evident that the environment has an impact on the handwriting. In non-identical twins, a significant difference is observed in the handwriting pattern of the twins both reared together and apart.

Comparing the handwriting of identical and non-identical twins reared together, it is observed that similarity is more in identical twins than non-identical. But it may be emphasized that heredity would simply make the resemblance in general, not in detail. However, handwriting is considered as a mental trait, which involves learning, and any detailed resemblance is due to the added influence of similar training. The detailed similarity in handwriting of identical twins when reared together, but not when reared apart, must, therefore, visualize in ascribing detailed mental resemblances of other kinds to heredity in the case of identical twins. It is, therefore, inferred that the hand-writing is dependant on heredity, environment, learning and training.

Intelligence Scoring of Twins

Intelligence scoring is another parameter of the study for which Raven's pictorial booklet (Fig. 5.25) was used. It is found that the coefficient correlation value of intelligence scoring among 40 pairs of identical twins reared together is 0.9411 whereas it is 0.896 for 39 pairs of non-identical twins reared together and 0.638 for 10 pairs of non-identical twins reared apart. However, the value for the nine pairs of identical twins reared apart is 0.9116, which is intermediate between identical and non-identical reared together. This shows that despite their separate environment, the intelligence of identical twins is more similar than non-identical.

Hence, it is evident that the environment in itself does not account for the greater similarity in the intelligence of identical twins as compared with non-identical. The greater similarity in question must be due to greater similarity of heredity. However, the environment does account for part of the similarity of identical twins. Otherwise those reared together would not be more similar than those reared apart. From this it is concluded that differences in the intelligence of people are due to partly heredity, and partly environment. However, even though two environments are somewhat different (as those of twins reared apart), both might still be "normal" in the sense that both allow for normal mental development, and the twin studies show that when the environment is normal, exceptional intelligence depends largely on exceptional heredity.

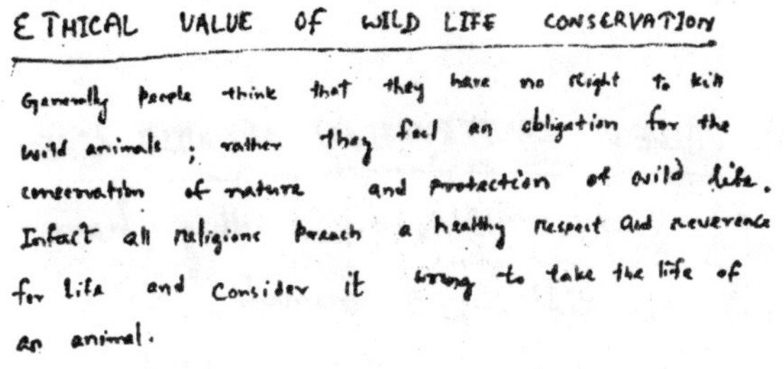

ETHICAL VALUE OF WILD LIFE CONSERVATION

Generally people think that they have no Right to kill wild animals ; rather they feel an obligation for the conservation of nature and protection of wild life. Infact all religions preach a healthy respect and reverence for life and consider it wrong to take the life of an animal.

First member.

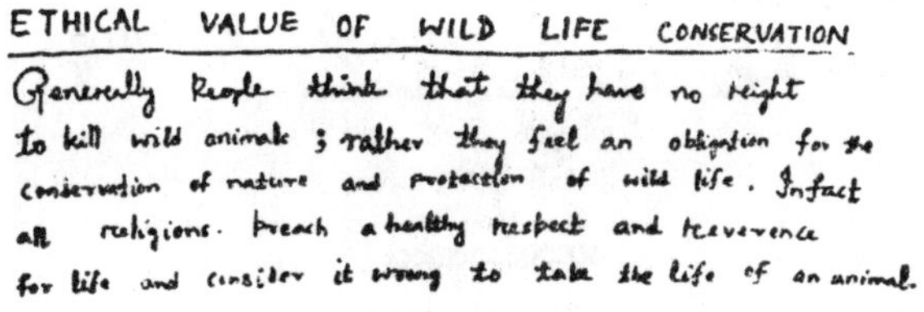

ETHICAL VALUE OF WILD LIFE CONSERVATION

Generally People think that they have no right to kill wild animals ; rather they feel an obligation for the conservation of nature and protection of wild life. Infact all religions preach a healthy respect and reverence for life and consider it wrong to take the life of an animal.

Second member.

Fig. 3.21. A sample page of handwriting of identical twins reared together.

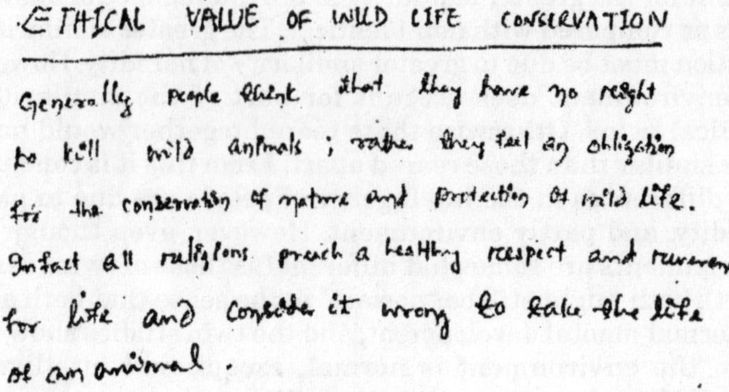

ETHICAL VALUE OF WILD LIFE CONSERVATION

Generally people think that they have no reight
to kill wild animals ; rather they feel an obligation
for the conservation of nature and protection of wild life.
Infact all religions preach a healthy respect and reverence
for life and consider it wrong to take the life
of an animal.

First member.

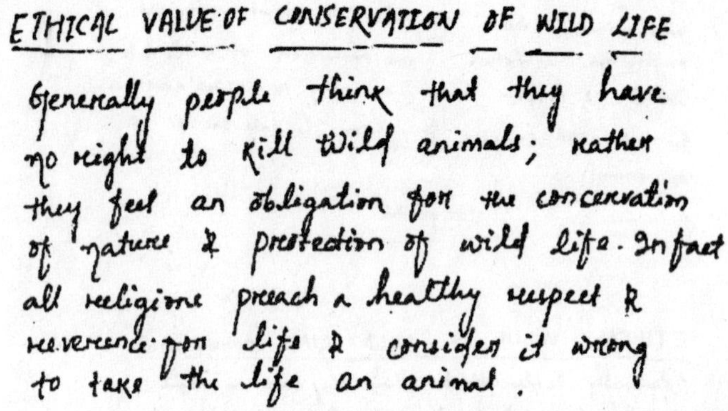

ETHICAL VALUE OF CONSERVATION OF WILD LIFE

Generally people think that they have
no right to kill wild animals; rather
they feel an obligation for the concervation
of nature & protection of wild life. Infact
all religions preach a healthy respect &
reverence for life & consider it wrong
to take the life an animal.

Second member.

Fig. 3.22. A sample page of handwriting of identical twins reared apart.

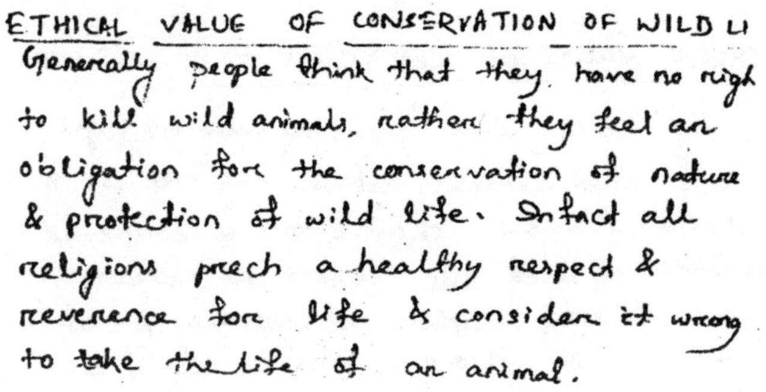

ETHICAL VALUE OF CONSERVATION OF WILD LI

Generally people think that they have no righ to kill wild animals, rather they feel an obligation for the conservation of nature & protection of wild life. Infact all religions prech a healthy respect & reverence for life & consider it wrong to take the life of an animal.

First member.

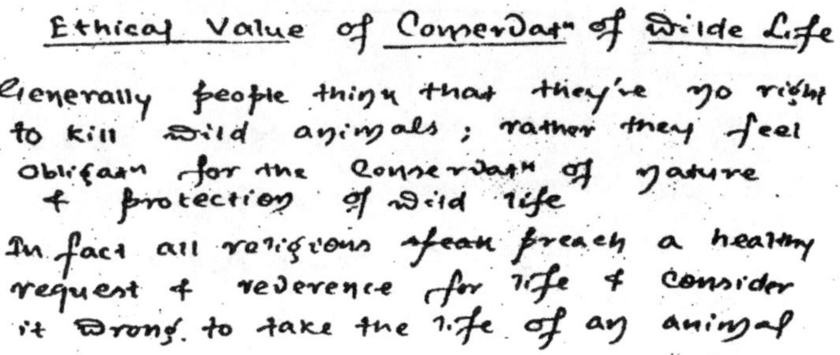

Ethical Value of Conservatn of Wilde Life

Generally people thinx that they've no right to kill wild animals; rather they feel obligatn for the conservatn of nature & protection of wild life

In fact all religions speak preach a healthy request & reverence for life & consider it wrong to take the life of an animal

Second member.

Fig. 3.23. A sample page of handwriting of Non- identical twins reared together.

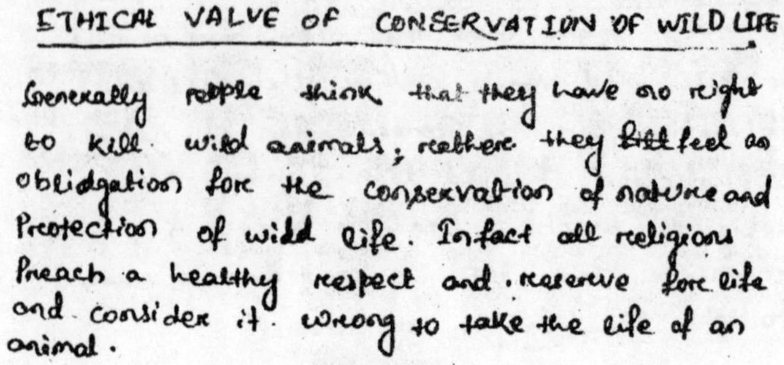

ETHICAL VALVE OF CONSERVATION OF WILD LIFE

Generally people think that they have no right to kill wild animals, rathere they ~~litt~~ feel a obligation for the conservation of nature and Protection of wild life. Infact all religions Preach a healthy respect and reserve for life and consider it wrong to take the life of an animal.

First member .

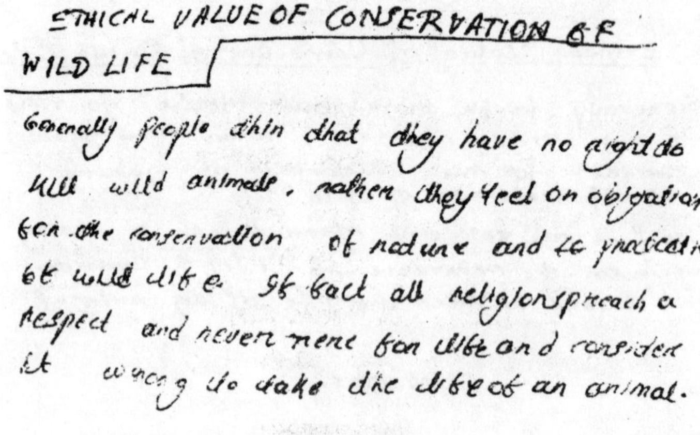

ETHICAL VALUE OF CONSERVATION OF WILD LIFE

Generally people thin that they have no right to kill wild animals, rather they feel on obligation for the conservation of nature and to protection of wild life. If fact all religions preach a respect and never rence for life and consider it wrong to take the life of an animal.

Second member .

Fig. 3.24.Handwriting of non-identical twins reared apart.

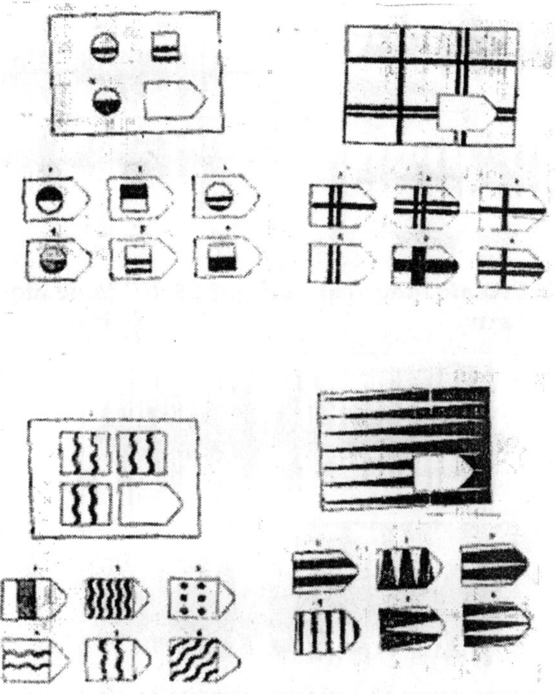

Fig. 3.25. A sample page of Raven's intelligence scoring booklet.

Pitch of the Voice of Twins

In relation to the pitch of the voice of the non-identical twins, 60 per cent of the male twins and 45.45 per cent of female twins show similarity and 100 per cent of male and female twins show dissimilarity. In toto, 27.45 per cent of non-identical twins show similarity in their pitch of the voice. With respect to the pitch of the voice of identical twins, 58.82 per cent of male and 83.33 per cent of female twins show more similarity. 20.58 per cent male twins and 5.55 per cent female twins show similarity and 20.58 per cent male twins and 11.11 per cent female twins show less similarity in their pitch of the voice. But in toto, 67.30, 15.38 and 17.30 per cent of identical twins show more similarity, similarity and less similarity respectively in their pitch of the voice. Comparing the pitch of the voice of identical and non-identical twins, it is inferred that the voice is dependant on heredity (Figs.5.26. (*i-v*)).

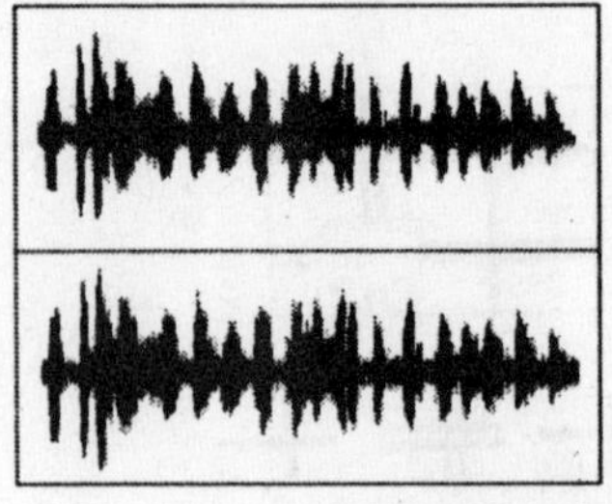

Fig. 3.26. (*i*) Female identical twins.

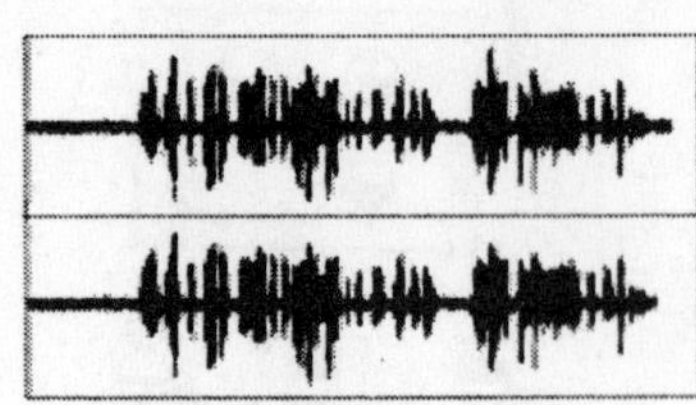

Fig. 3.26. (*ii*) Male identical twins.

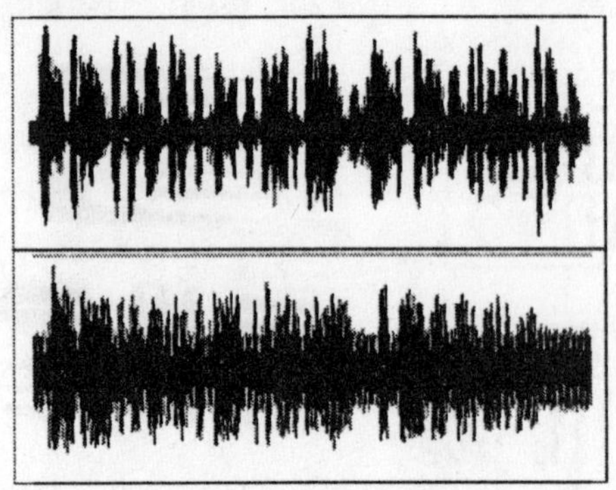

Fig. 3.26. (*iii*) Male female non-identical twins.

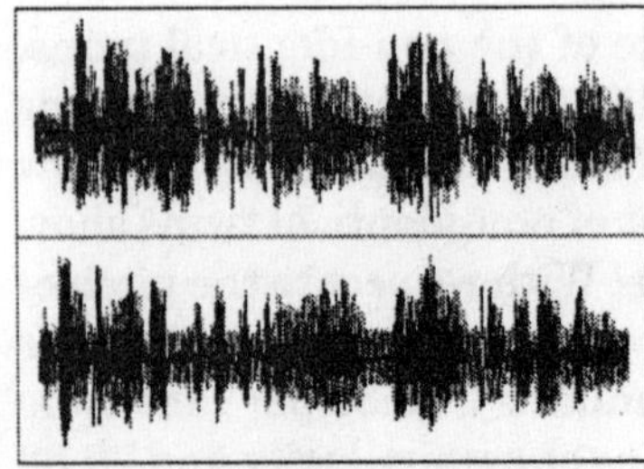

Fig. 3.26. (*iv*) Female non-identical twins.

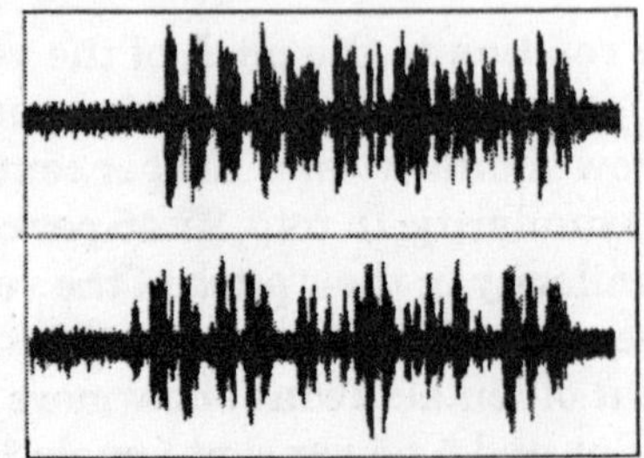

Fig. 3.26.(*v*) Male non-identical twins.

Fig. 5.26. Comparison of pitch of the voice between the members of identical twins and in non-identical twins(a single diagrammatic sample).

4
Hairwhorl of Twins

Hairs are specialized, elongated thread-like and cylindrical outgrowths of the mammalian epidermis. Some of the notable uses of hairs are :

(*i*) imparting colour and camouflage,

(*ii*) forming an insulating sheath by trapping air which presents undue loss of heat from the body,

(*iii*) forming a kind of portable blanket infurred for survival in cold,

(*iv*) acting as tactile receptors (vibrissae),

(*v*) preventing dust particles by eye lashes and hairs in and around nostrils and

(*vi*) helping in getting rid of flies and other harmful insects by long tail hairs of cow or horse.

The hair provides evidences for individual as well as racial identification. Although the general structure of all human hairs is similar, some variation occurs among ethnic groups. The hair of scalp has been studied more exhaustively than hair of other regions of the body, because of its greater abundance and accessibility. Its length, thickness, colour and texture are gross characteristics seen with the naked eye, which are useful to anthropologists in distinguishing ethnological group. However, hairwhorl is a characteristic feature of all human beings. Whorl is defined as circular arrangement of any object. The description of the whorl with respect to spinning, fingerprint and flowers are available in literatures but literature is silent on scientific account of hairwhorl of human beings.

Hairwhorl, formed by hair is present on the rearside (crown) of the head since the birth of human babies. This hairwhorl is usually singular in each individual which may be of clockwise or anti-clockwise orientation. But rarely the structure, number and orientation of this hairwhorl show certain variation. The hairwhorl may also be noticed in paired or fused form. In paired condition, similar hairwhorl or dissimilar hairwhorls are present and they may be located closely or distantly or unitedly.

In twins members nine types of hairwhorls are identified such as clockwise, anti-clockwise, paired clockwise distantly located, paired clockwise closely located, paired anti-clock and clockwise distantly located, paired anti-clock and clockwise closely located, paired clock and anti-clockwise distantly located, paired clock and anticlockwise united and paired anti-clock and clockwise united and which are located on the crown of the heads (Table 4.1 and Figs. 4.1 and 4.2). Analysis shows percentage of concordance in hairwhorls of identical twins to be more than non-identical twins. That idicates more genetic identicality in identical twins than that of non-identical twins. The percentage of clockwise hairwhorls are more in both identical and non-identical twins. But paired hairwhorls are found rarely in both categories of twins. This hairwhorl seems to be significant to establish genetic identicality, inheritance and cleavage pattern during embryonic development.

Table 4.1. Types of hair whorls in twins

Sl. No.	*Hair whorl*	*Location on head*	*Figures of hair whorls*
1.	Clockwise	Front or rear or both	
2.	Anti-clockwise	Front or rear or both	
3.	Paired clockwise distantly located	Front or rear or both	
4.	Paired clockwise closely located	Front or rear or both	
5.	Paired anti-clock wise and clockwise distantly located	Front or rear or both	

Sl. No.	*Hair whorl*	*Location on head*	*Figures of hair whorls*
6.	Paired anti-clock and clockwise closely located	Front or rear or both	
7.	Paired clockwise and anti-clockwise distantly located	Front or rear or both	
8.	Paired clock and anticlockwise united	Front or rear or both	
9.	Paired anticlock and clockwise united	Front or rear or both	

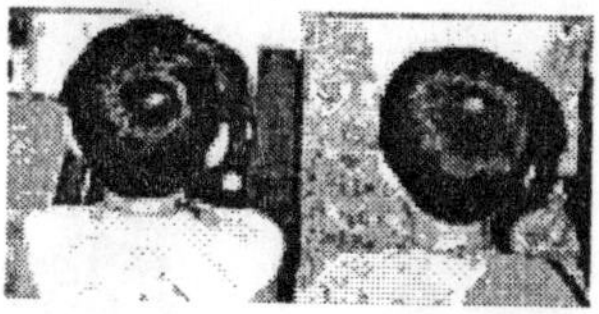

(*a*)

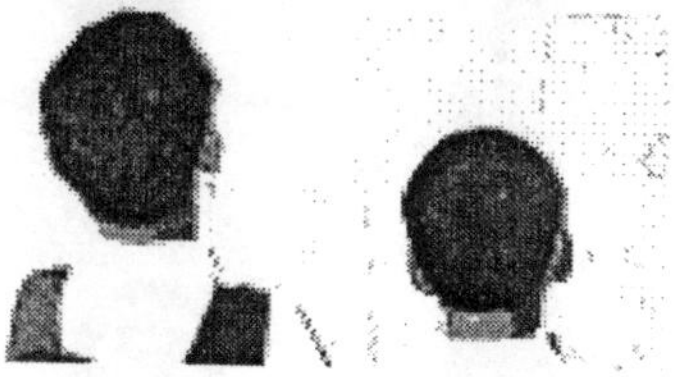

(*b*)

(*c*)

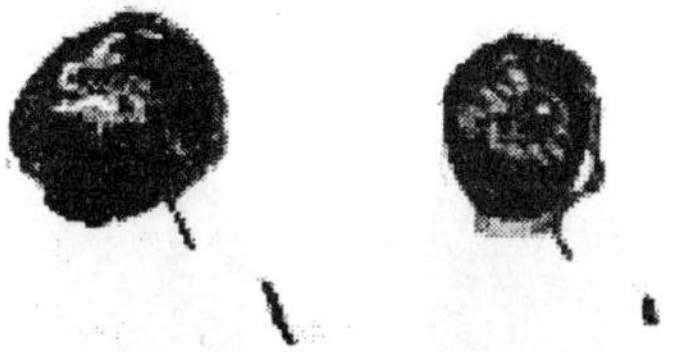

(*d*)

(*e*)

(*f*)

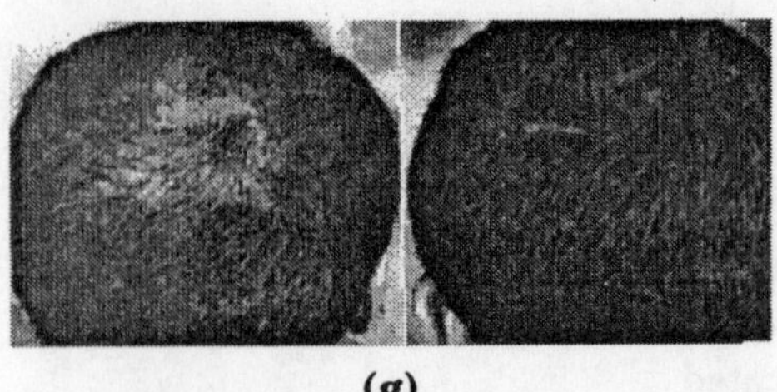

(*g*)

Fig. 4.1. Hairwhorls is non-identical twins.

(*a*) Clockwise and clockwise; (*b*) Anti-clockwise and anti-clockwise; (*c*) Anti-clockwise and clockwise; (*d*) Paired clockwise distantly located and a singular clockwise; (*e*) Clockwise and paired distantly located clock and anticlockwise, (*f*) Paired anti-clock and clockwise united and clockwise; (*g*) Paired clock and anti-clockwise united and paired clockwise closely located.

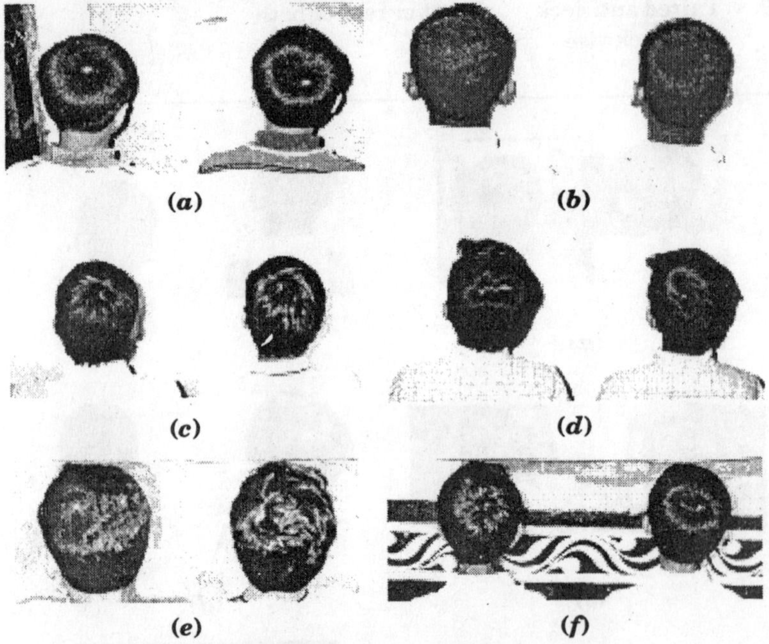

(*a*) (*b*) (*c*) (*d*) (*e*) (*f*)

Fig. 4.2 Hairwhorls in identical twins.

(*a*) Clockwise and clockwise; (*b*) Anti-clockwise and anti-clockwise; (*c*) Anti-clockwise and clockwise; (*d*) Paired clockwise distantly located and clockwise; (*e*) Paired anti-clock and clockwise distantly located and a singular anti-clockwise, (*f*) A singular clockwise and paired anticlock and clockwise closely located.

5

Dactylography of Twins

History of Fingerprint Identification

Dactylography (Gr. *daktylos*-finger; *graphein*-to write) is the scientific study of fingerprints. The identification from fingerprint is even now a subject of mystery to the general public. It is not known who and when first used fingerprint for the purpose of identification, but the study of hands and uses of fingerprints can be traced back to the very earliest form of civilization. The inner surface of human hand termed as palmer or thenar surface (Fig. 5.1) and the under surface of the foot, known as plantar surface, are covered with a thick layer of skin which shows an alternate arrangement of elevations and depressions. The elevations are known as ridges and depressions as furrows. Besides ridges and furrows, crease marks are also seen which are of two kinds such as permanent and temporary. The former are formed in the embryo by folding of skin and are of interest to the palmist. These are line of head, line of heart, line of life etc. and include interphalangeal space lines and basal phalangeal lines (phalanx marks).

Besides the phalanx mark, another mark is present transversely between the phalanx marks in the reported case. This mark is observed to be similar to phalanx mark and is named as intraphalangeal marks basing on its location (Fig. 5.1). These marks are observed since their birth. But the temporary creases appear as a result of skin diseases, deformity, illness, tenderness and old age. Permanent creases which are rather prominent help in comparing the palm prints. In an inked impression of fingers, the print shows ridges and furrows which are represented by alternate black and white lines run side by side respectively whereas temporary creases run crosswise or randomly in any direction against the lines of ridges and furrows. These show inward folding of skin surface which passes through ridges and furrows, but there being no dislocation in the continuity of ridges and furrows (Fig. 5.1).

The finger balls of all human beings are unique in having different varieties of specialized form of designs called as patterns. The print of the finger balls shows different patterns such as loop, whorl, arch and composite. It differs significantly from person to person. In few cases, when prints do not show any difference in patterns then, they exhibit difference in ridge numbers, which are present between the core and delta. The comparison of fingerprints of different persons possesses remarkable difference in their prints.

Fingerprints along with palm prints known as *'panja'* were used for some centuries in India. Dr Henry Faulds who came to Darjeeling for a short time in 1872 as a medical missionary under the Church of Scotland went to Japan later and while working at the Tsukiji Hospital in Tokyo in 1877 noticed some finger marks on some Japanese pottery which he composed with the finger impressions of people. In 1880 he published a report of his experiments showing methods of taking fingerprints with printer's ink and discussing the various directions of study to which fingerprints may lead. He was the first man to give the idea of tracing a criminal from a chance impression.

Sir William Herschel, who was collector of Hoogly district in Bengal had been interested in fingerprints since 1858. The first print he made was on a contract with a Bengali contractor named Rajyadhar Konai. On the contract form he took the fingerprints along with the palm print of the right hand of the signer so as to frighten him from denying his formal act. Today this so-called 'historic contract' is still preserved.

Sir Herschel extensively used fingerprints officially in the province of Bengal, for identification between the years 1858-1880. He also submitted a report to the Government advocating the adoption of fingerprints through out the province incase of prisoners, but his report was not adequately attended to. Sir

William Herschel did not publish anything on the use of fingerprints before Faulds.

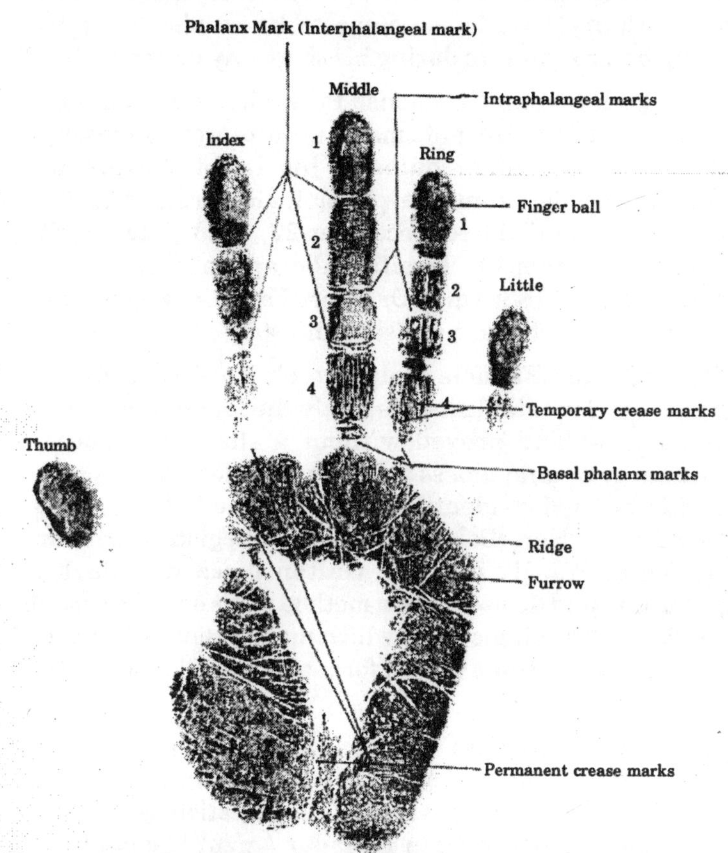

Fig. 5.1. Palm and fingerprints showing ridges, furrows and temporary and permanent creases.

In Bengal, the exact time period of *"Tip sahi"* (signature by impression) use is not known exactly. According to Herschel this *'tip sahi'* was put as mere formality by illiterate person who could not sign even a 'cross' or 'caste mark'. Again palm prints were used in India long before *Rajyadham Konais* handprint was made by Sir William Herschel. But he did not mention any where in his publication about the of palm prints in India.

According to many authors Herschel must have hit upon the taking of his finger pad observation from the known use of fingerprints in India in contracts or customs. It is not unlikely that Dr Henry Faulds too came to know about the use of fingerprint for signature during his short stay in Bengal in 1872.

In 1823 Dr Johan Evangelish Purkinje, Prof. of Anatomy of the University of Breslan, made an attempt to classify the patterns on the finger bulbs in his Latin thesis entitled "commentalio de examinee physiologico organ versus etc. systematic cutaur" dated December 22, 1823. After Purkinje, no further system of classification appears to have been undertaken until 1888 when Dr Henry Faulds devised a method, an outline of which was published in 1905.

In 1890 Sir Francis Galton took up the study of the fingerprints. He established securely the foundation of science of fingerprint and proved without a shadow of doubt the individuality or uniqueness and permanence of fingerprint as a scientific method of identification with the help of materials collected by Sir Herschel and special Sub-Registrar Sri Ramgati Banerjee of Hoogly district. Galton discarded Purkinje's classification and devised a new method of his own for describing and indexing the single print which made it somewhat possible to search a single thumb print form a large collection of thumb prints.

Sir Edward Richard Henry, Inspector General of Police, lower Bengal conducted a series of experiments and by 1896, he was able to develop a new system of classification of fingerprints. Two Bengalee officers Khan Bahadur Azizul Haque, the then officer-in-charge of the Anthropometrics Bureau and Rai Bahadur Hem Chandra Bose, an officer of the Bureau (unfortunately their names have not been mentioned in the book by Sir Edward Richard Henry) played very important part in the development of the system. It was left to peculiar Bengalee genius to invent sub-classification of types and a system of sorting pegion holes which finally rendered it as easy to turn up a man's fingerprint slip as it was to turn up Bertillon's anthropometrics cards in the identification Bureau. This system was proved so satisfactory in 1896 that an application was made to the

Government of India for the appointment of a committee to report up on the system.

The Committee consisting of Mr C. Strahan, R.E. Surveyor General of India and Mr Alex Pedler, F.R.S. Principal, Presidency College, Calcutta (Kolkata) examined both anthropometric system and Henry's system of classification of fingerprints and recommended for the adoption of fingerprint system as means of identification of the habitual criminals in place of anthropometrics system in March, 1897. Then the first Finger Print Bureau in the world was officially established in Calcutta in June, 1897. In a short time, the Government of India were so fully convinced of the effectiveness of the new system and of authenticity of the result, that in 1899 the Indian legislature passed a special Act amending the law of evidence to the extent of declaring relevant the testimony of those who had become proficient in fingerprint decipherments.

Fingerprint Pattern and Ridge Characteristics

The ridges on the finger balls form various 'patterns', which are capable of classification. Sir Francis Galton sorted nine patterns classified by Dr Evanjelist Purkinje into three classes namely arch, loop and whorl.

Pattern

Whenever an interspace is left between the boundaries of a different system of ridges, it is fitted by a small system of its own, which will have some characteristic shape and is called a 'pattern'. Sir Edward Richard Henry modified Galton's A.L.W.system and classifed them into four groups (Fig. 5.1) such as :

(*i*) arch,

(*ii*) loop,

(*iii*) whorl and

(*iv*) composite (Fig. 5.2).

His system of classification is the most efficient and is now in use through out the world. The classification of patterns depends on the formation of fixed points such as (*i*) outer terminus in delta and (*ii*) inner terminus in core.

Delta

The word 'delta' is the fourth letter of the Greek alphabet and corresponds to the English letter 'D'. The Greek letter is triangular in shape; hence the word is applied to many things shaped like triangles such as island at the mouth of rivers. There is great resemblance between the delta in geography and delta in fingerprint patterns though in some cases of fingerprint the delta may not be triangular shaped. When a ridge bifurcates, and the two arms of the bifurcating ridge diverge causing an interspace within which the pattern lies, the point of bifurcation is called "delta".

When two adjacent ridges, which had run side by side to some extent, diverge causing an interspace within which the pattern lies, the triangular plot formed by the two diverging ridges and the first ridge in front of them within the interspace as the base is called delta [Fig. 5.4.(*i-iv*)]. When delta is formed by bifurcation of a single ridge, the point of bifurcation is the 'outer terminus'. When delta is formed by the divergence of two ridges which run side by side, the first ridge in front of the place where divergence begins, even it be a mere point, whether it is independent or is springs form the divergent ridges is the outer terminus and represented as O.T. [Fig. 5.4. (*v-xx*)].

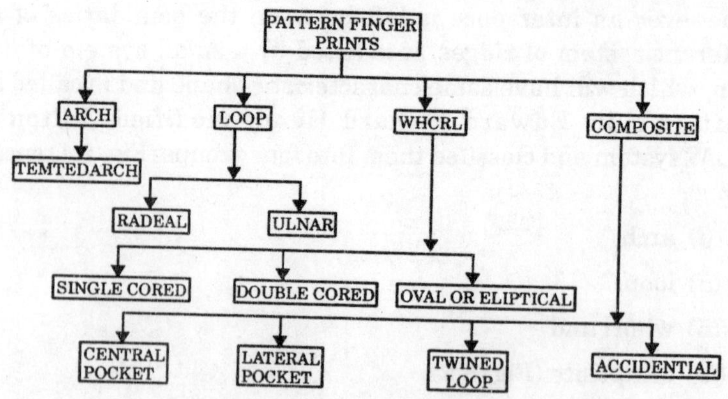

Fig. 5.2. Outline of classification of fingerprint.

Core

The core means the central part of a pattern. The type of the core varies according to the pattern. The core of a loop may consist of staple or an even or uneven number of ridges not joined together, which are called 'rod' (Fig. 5.4.vii-xiii), symbolized as r. In case of staple being a core, the point on its s shoulder farthest from the delta is the 'inner terminus'. When the core consists of even number of ridges (rods) the two central ridges are considered as joined at their summit by an imaginary neck and point farthest from delta on the shoulder is the inner-terminus.

In whorl the center of the first ring, elliptical or circular, is the point of core. In case of spiral, the point from which the spiral begins to revolve is the point of core (Fig. 5.4.iv). Sir Edward Richard Henry classified patterns into four groups such as :

(*i*) arch,

(*ii*) loop,

(*iii*) whorl and

(*iv*) composite which are represented by the symbol A,L, W and C respectively (Fig. 5.2).

Arch

An arch is a pattern in which the ridges extend from one side of the finger to the other without turning back, rising slightly at the center, where the curvature looks like an arch. There is ordinarily no delta (Fig.5.3.i). However, ridges near the middle of the base have an upward thrust arranging themselves as it were on both sides of a spine or axis towards which adjoining ridges converge. The ridges,thus, converging give to the pattern the appearance of a tent in out line. This type of arch is called a 'tented arch' (Fig. 5.3.ii).

Loop

A loop is a pattern in which one or more ridges recurve, that is, run back on their previous course, making a half turn or more around the core, having a delta and at least one ridge intervening between the inner and outer termini. There are two kinds of loop such as (1) ulnar and (2) radial.

Ulnar

An ulnar loop is so called, because the ridges terminate in the direction of ulnar bone of the fore arm, in other words, the ridges slant towards right in case of right hand fingers and towards left in case of left hand fingers (Fig. 5.3.iii).

Radial

A radial loop is so called, because the ridges terminate in the direction of the radial bone of fore arm, that is the ridges slant towards left in case of right hand fingers and to the right in case of left hand fingers (Fig. 5.3.iv).

Whorl

The word whorl in the English language means a number of leaves in a circle round the stem; but in relation to recurring ridges, it represents a type of pattern in which the extent of curvature around the core is one complete twist or circuit, and there are two deltas. There may be one or more ridges making a turn at least one complete circuit. Broadly, they are either single cored (Fig. 5.3.v) or double cored, having two cores (Fig. 5.3.vi) or elongated in their central portion, which is oval or elliptical (Fig. 5.3.vii). In small central circular whorls, the innermost ring is the core and in spiral whorls the point from where the core ridge begins to revolve is taken as the point of core.

Composite

According to the English dictionary composite means compound, that is, not simple. The same meaning applies to this word when used in the science of fingerprints. All those impressions which show a combination of two or more ridge designs with two or more deltas are grouped under composites. There are four distinct varieties of composites namely,

1. central pocket loop,
2. lateral pocket loop,
3. twinned loop, and
4. accidental. Like whorls these are also sub-divided by ridge tracing, but when there are three deltas the two outer one only are used for this purpose.

(*i*) **1 Central Pocket Loop:** In central pocket loop pattern, there is a combination of loop and whorl (Fig. 5.3.viii). The whorl formation lies in the core region of the pattern in the form of a pocket, which is formed when one or more ridges immediately near the core deviate in course from the general course of the other recurring ridges. The majority of ridges, thus,flow in a hairpin like manner. In this case, two deltas exist. It is not, however, essential that the whorl like appearance must show at least one complete circuit. Whenever there is an obstruction at right angles to the line of exit or flow of ridges, the pattern is taken as a central pocket loop. However, when the ridges or ridges about the core do not recurve, but meet this line at an acute angle, the design is included under the loop type of pattern. The line of exit is an imaginary line drawn between the inner delta and the centre of the innermost looping ridge.

(*ii*) **2 Lateral Pocket Loop:** Lateral pocket loop is a type of pattern, which consists of two inter clasped loops, one overlapping the other and forming a definite side pocket (Fig. V.3.ix). There are two deltas and two cores and the ridges containing the points of core of the two loops, when traced, flow out of the pattern undivided by either of the deltas, that is, emerge from the same side of either delta.

(*iii*) **2 Twinned Loops:** In twinned loops type, there are two well-defined loops. One loop, more or less, surrounds the other. There are two deltas and two cores like the lateral pocket loop, but in this type of configuration the ridges containing the points of core of the two loops, when traced, flow out on the opposite side of the interposed deltas (Fig. 5.3.x).

(*iv*) **4 Accidental:** Some composite patterns are called accidental as they rarely occur, such as loop-by-loop, whorl resting on loop, loop resting on whorl, whorl resting on whorl, and arch with pocket. According to Henry loop by loop is a lateral pocket. Bose defines it as an accidental one differentiating it from lateral pocket. In loop by loop, both the deltas are on one side, whereas in lateral pocket, one delta is on the right and one on the left (Fig. 5.3.xi).

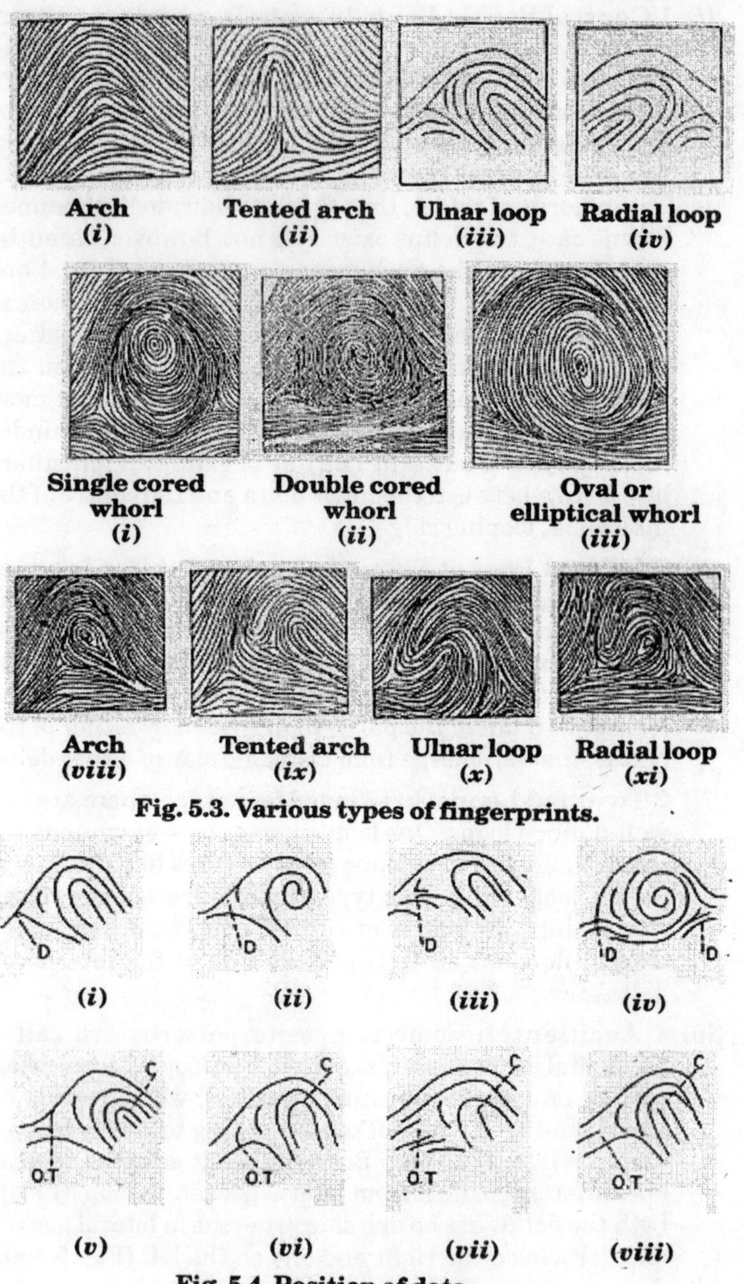

Fig. 5.3. Various types of fingerprints.

Fig. 5.4. Position of data.............

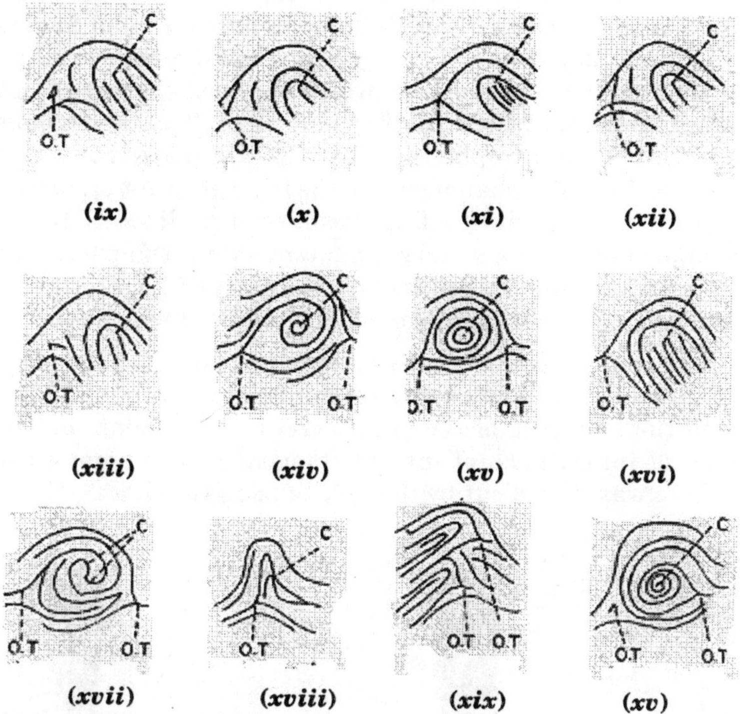

Fig. 5.4. Position of delta, core and outer terminus in different types of fingerprints.
D – Represents to delta.
C – Represents to core.
O.T – Represents to outer terminus.

In dactylography the percentage of concordance in the respective hands of identical twins has been observed to be more than non-identical which implies that the finger patterns of identical twins are more alike than non-identical. The patterns, which have been studied, are loop, whorl, arch and composite. This signifies a greater influence of genetic similarity on the concordant appearance as such traits have strong genetic components in identical than non-identical twins.

Variation in Human Fingerprint Trait and Inheritance

No two fingerprints of human beings are equal or identical. Even fingerprints of both the hands of same individual are not identical. The phalanx marks are almost clearly visible in all individuals, which are normally three in number in all fingers except the thumb, where it is only two. A case having variation in fingerprints is reported here which has been observed in an identical twin pair, their mother and niece. This anomalous fingerprints show intraphalangeal marks (Fig. 5.5) in the stated members. The intraphalangeal marks in fingerprint of human beings are a new finding in fingerprint science. Since no report is available on such anomalous phalanx mark, the particular investigation is undertaken to study the detailed fingerprints of a twin pair and their family members (mother and niece).

With respect to finger patterns and intraphalangeal marks, percentage of similarities and dissimilarities were found out among the members of the family and in their relation. The nature of inheritance of intraphalangeal marks among the members was traced out by the help of pedigree chart.

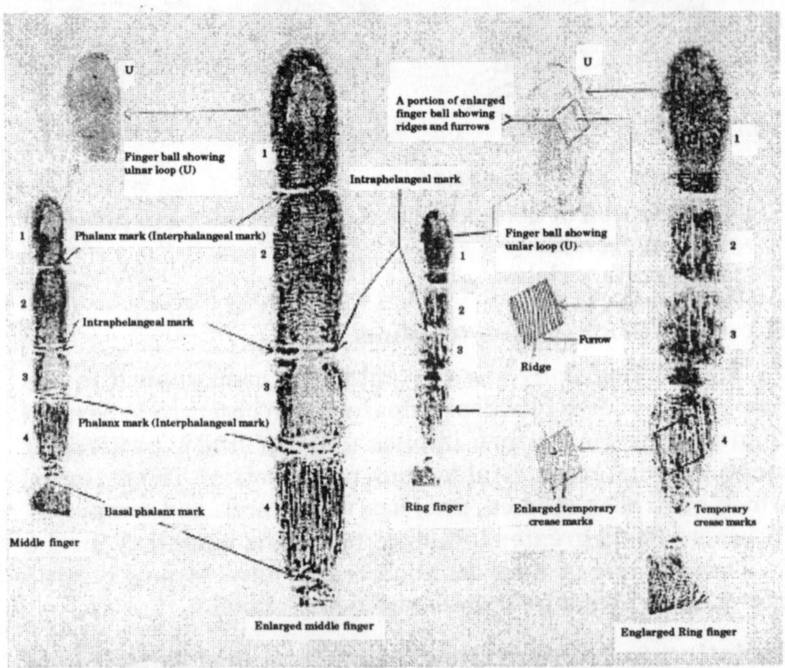

Fig. 5.5. Fingerprint of middle and ring finger showing ridges, furrows, temporary and permanent creases.

Fig. 5.6. Identical twin sisters Sabita and Namita.

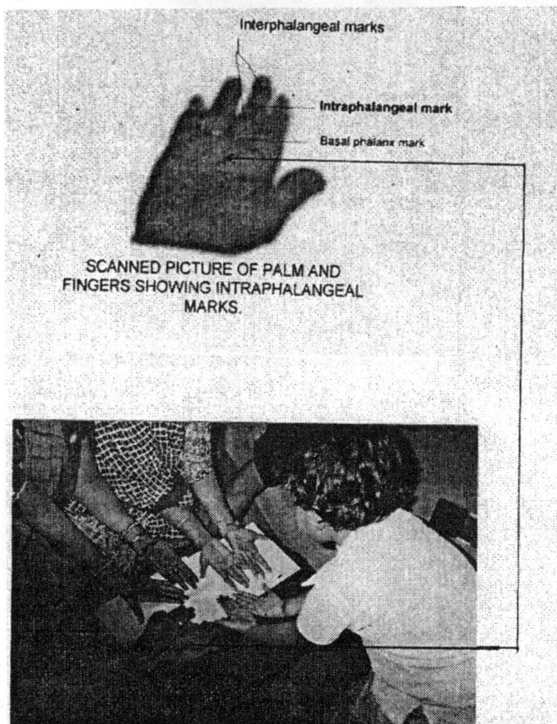

Fig. 5.7. Identical twin pairs, their mother and niece showing intraphalangeal marks.

Finger Pattern

The fingerprints of the members of the family show finger patterns like loop, whorl, arch and twinned loop. Their similarities in stated patterns in respective fingers among the four members have been analysed.

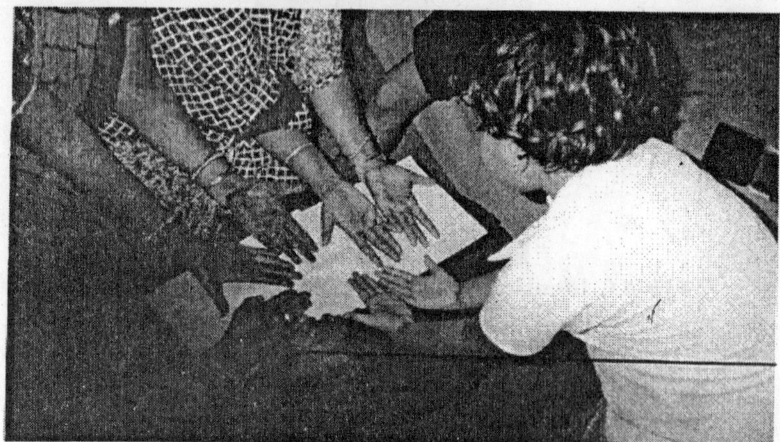

Fig. 5.8. Identical twin pairs, their mother and niece showing intraphalangeal marks (enlarged view).

The identical twin sisters Sabita and Namita, show 60 per cent of concordance (Tables 5.1 and 5.2), in fingerprint patterns of their respective right hands (Figs. 5.9, 5.10 and 5.13) and 80 per cent of concordance (Tables 5.1 and 5.2) in fingerprint patterns of their respective left hands' finger balls (Figs 5.9, 5.10 and 5.14). The twin sisters and their mother show 60 per cent concordance (Tables 5.1 and 5.2) in each of their respective right and left hands (Figs. 5.9, 5.10, 5.11, 5.13 and 5.14). The twin sisters and their niece show 40 per cent concordance (Tables 5.1 and 5.2) in each of their respective right and left hands (Figs. 5.9, 5.10, 5.12, 5.13 and 5.14). Further, 40 and 80 per cent concordance (Tables 5.1 and 5.2) are found in the right and left hands respectively in their mother and niece (Figs. 5.11, 5.12, 5.13 and 5.14). But 40 per cent concordance is recorded both in right and left hands (Tables 5.1 and 5.2) of all the four members (Figs. 5.9, 5.10, 5.11, 5.12, 5.13 and 5.14).

Intraphalangeal Mark

The twin sisters show similarity in inheritance of deep intraphalangeal mark similar to phalanx mark, present between the

Table 5.1 Fingerprint patterns of family members including a twin pair

Sl. No	Name	Right hand					Left hand				
		Thumb	Index	Middle	Ring	Little	Thumb	Index	Middle	Ring	Little
1	Asha Manjari Mohanty (Mother of twin pair)	U*	U	A**	U	U	U	U	U	U	U
2	Sabita Mohanty (Twin member)	U	TL***	A	U	U	TL	U	U	U	U
3	Namita Mohanty (Twin member)	U	U	U	U	U	TL	U	A	U	U
4	Dibyasha Mohanty (Niece of twin pair)	W****	W	U	U	U	U	U	U	W	U

*U-Ulnar loop ** A-Arch, ***TL – Twinned loop,****W- Whorl

Table 5.2. Percentage of concordance in finger pattern of family members

Sl.No.	*Members of a family*	*Right hand*	*Left hand*
1.	Twin sisters	60	80
2.	Mother and twin sisters	60	60
3.	Twin sisters and niece (Brother's daughter)	40	40
4.	Grand mother and grand daughter (Niece of twin pair)	40	80
5.	Among the four members	40	40

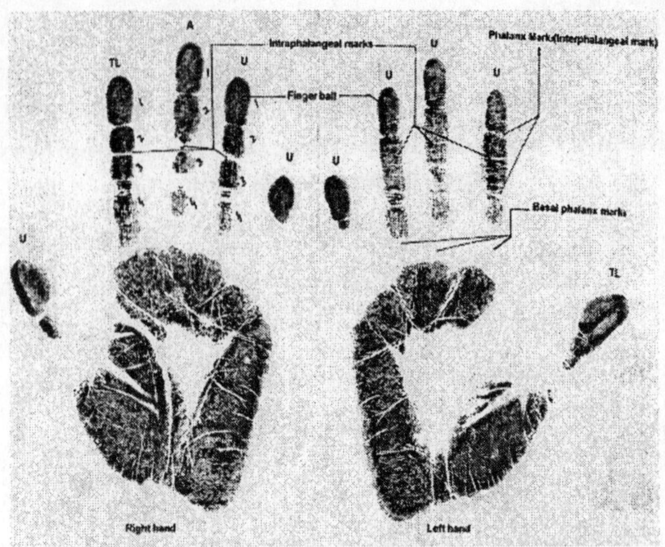

Fig. 5.9. Fingerprints of the one of the twins(Sabita Mohanty).

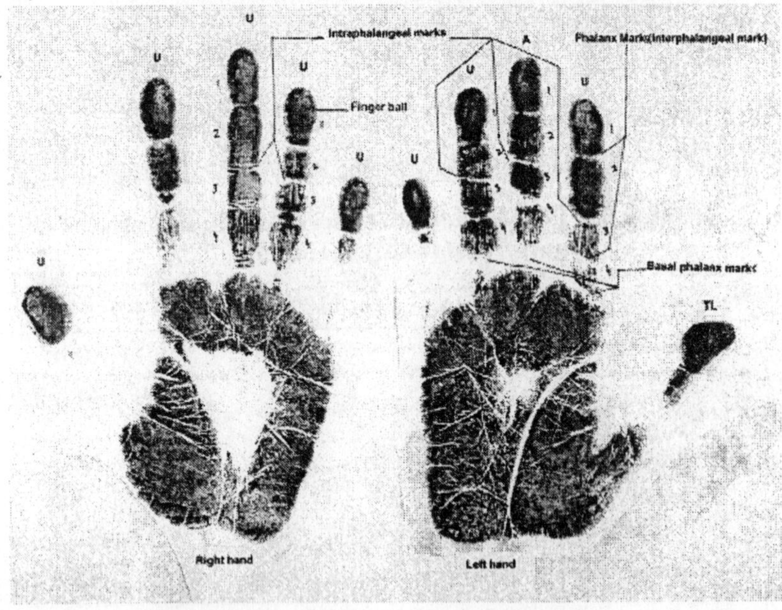

Fig. 5.10. Fingerprints of another twin (Namita Mohanty).

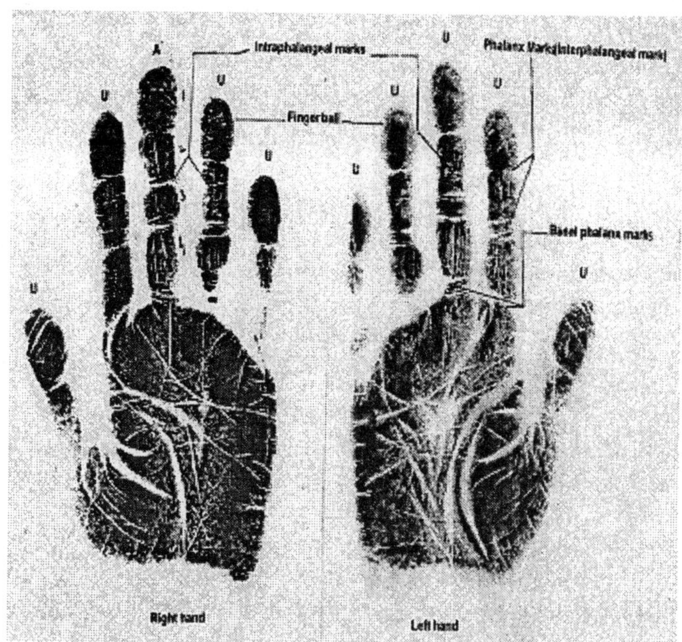

Fig. 5.11. Fingerprints of mother of the twins (Asha Manjari Mohanty).

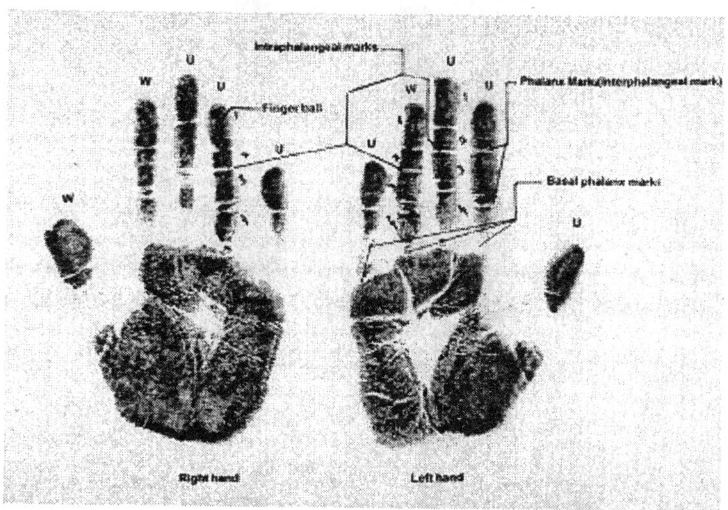

Fig. 5.12. Fingerprints of niece of the twins (Dibyasha Mohanty).

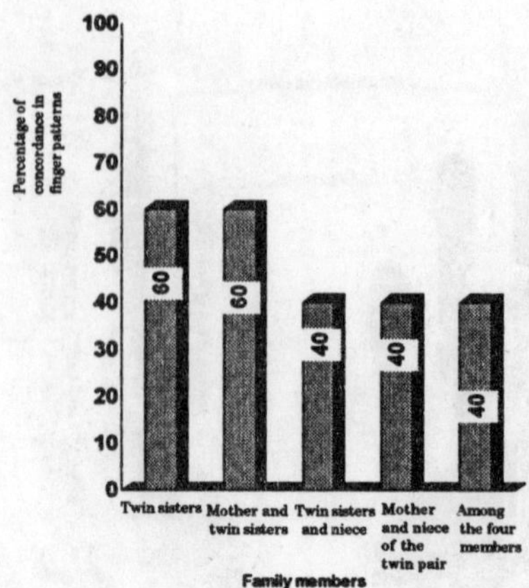

Fig. 5.13. Percentage of concordance in finger patterns of right hands of the family members.

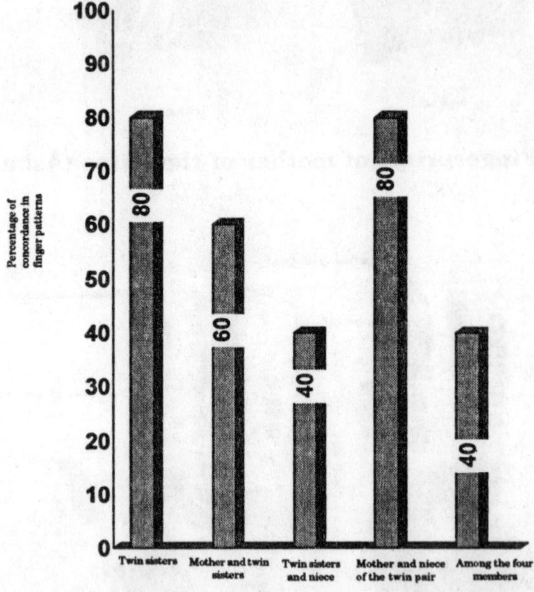

Fig. 5.14. Percentage of concordance in finger patterns of left hands of the family members.

phalanx marks on the ventral side of phalanges (Table 5.3 and Figs. 5.9 and 5.10). The ventral side of index,middle and ring fingers of both right and left hands of Sabita shows one deep intraphalangeal mark similar to phalanx mark, which is present between the two phalanx marks in each of this finger (Fig. 5.9). Besides, three phalanx marks in each finger of Namita (Fig. 5.10), the middle and ring fingers of right hand and index, middle and ring fingers of left hands show one deep intraphalangeal mark.

One deep intraphalangeal mark similar to phalanx mark and a faint intraphalangeal mark are observed in middle and ring fingers of right hand respectively. One faint intraphalangeal mark is also present in the middle finger of left hand of the mother of twin sisters, Asha Manjari (Table 5.3 and Fig. 5.11). Dibyasha, the grand daughter of Asha Manjari,possesses a deep intraphalangeal mark similar to phalanx mark. This is present between the phalanx marks on the ventral side of ring finger of right hand and middle and ring fingers of the left hand (Table 5.3 and Fig. 5.12).

Percentage of Similarity in Intraphalangeal Marks

Twin sisters show 67 and 100 per cent of concordance in intraphalangeal marks in their fingers of right and left hands respectively (Table 5.4 and Figs. 5.15 and 5.16). The mother and twin sisters show 67 and 33 per cent of concordance in intraphalangeal marks in their right and left hand fingers respectively (Table 5.4 and Figs. 5.15 and 5.16). Twin sisters and niece show 33 and 67 per cent of concordance in intraphalangeal marks in their right and left hand fingers respectively (Table 5.4 and Figs. 5.15 and 5.16). However, mother and niece of the twin pair exhibit 33 per cent of concordance in intraphalangeal marks in both of their right and left hand fingers (Table 5.4 and Figs. 5.15 and 5.16). Further, 33 per cent of concordance is found in both of their right and left hand fingers among all the four members (Table 5.4 and Figs. 5.15 and 5.16).

members including a twin pair

Sl. No	Name	Right hand					Left hand				
		Thumb	Index	Middle	Ring	Little	Thumb	Index	Middle	Ring	Little
1	Asha Manjari Mohanty (Mother of twin pair)	Absent	Absent	Intraphalangeal mark (deep)	Intraphalangeal mark (faint)	Absent	Absent	Absent	Intraphal-ngeal mark (faint)	Absent	Absent
2	Sabita Mohanty (Twin member)	Absent	Intraphalangeal mark (deep)	Intraphalangeal mark (deep)	Intraphalangeal mark (deep)	Absent	Absent	Intraphalangeal mark (deep)	Intraphal-angeal mark (deep)	Intraphalangeal mark (deep)	Absent
3	Namita Mohanty (Twin member)	Absent	Absent	Intraphalangeal mark (deep)	Intraphalangeal mark (deep)	Absent	Absent	Intraphalangeal mark (deep)	Intraphalangeal mark (deep)	Intraphalangeal mark (deep)	Absent
4	Dibyasha Mohanty (Niece of twin pair)	Absent	Absent	Absent	Intraphalangeal mark (deep)	Absent	Absent	Absent	Intraphalangeal mark (deep)	Intraphalangeal mark (deep)	Absent

Table 5.4. Percentage of intraphalangeal marks on the similar position of the fingers of family members

Sl.No	*Family members*	*Right hand*	*Left hand*
1	Twin sisters	67	100
2	Mother and twin sisters	67	33
3	Twin sisters and niece	33	67
4	Grand mother (mother of twin pair) and grand daughter (niece of twin pair)	33	33
5	Among four members (Twin sisters, mother and niece)	33	33

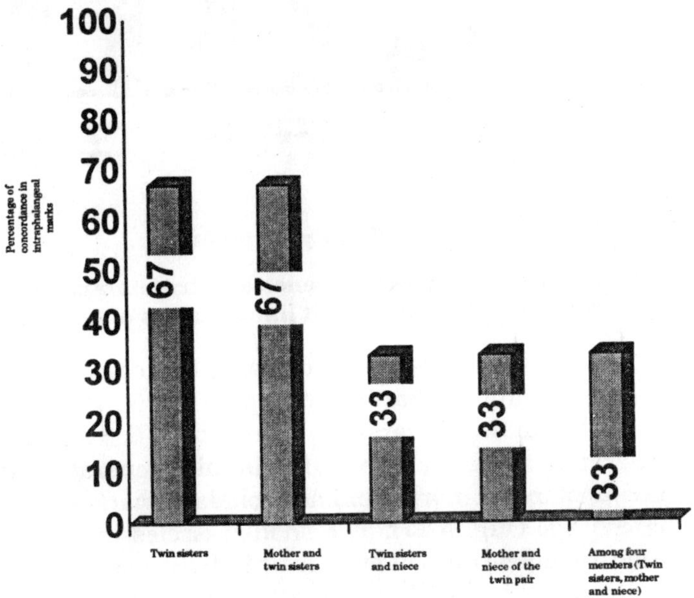

Family members

Fig. 5.15. Percentage of presence of intraphalangeal marks on the similar position of the right hand fingers of family members.

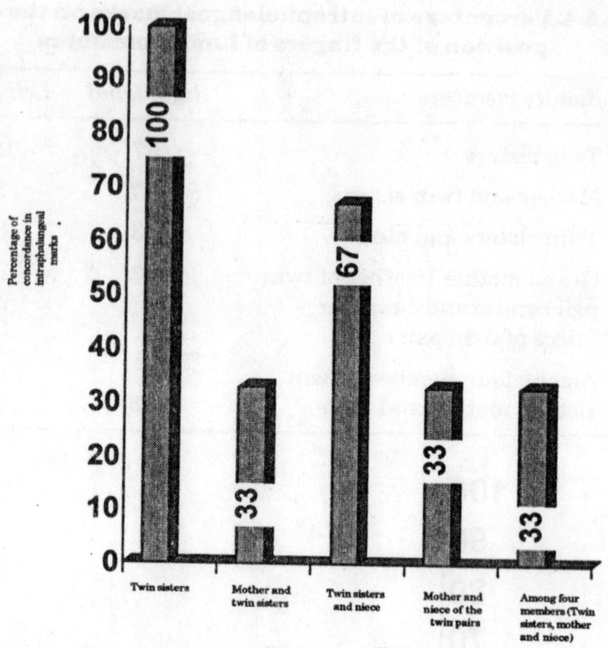

Family members

Fig. 5.16. Percentage of presence of intraphalangeal marks on the similar position of the left hand fingers of family members.

Basing on the presence of intraphalangeal marks on the fingers, a pedigree chart is developed (Fig. 5.17). The mother of twins (I-2) in first generation, twin sisters (II-3 and 4) in second generation and their niece (III-1) in third generation show such intraphalangeal marks similar to phalanx mark with a different percentage (Fig. 5.17). The shaded circles in pedigree chart indicate appearance of such trait in twin sisters, their mother and niece (Fig. 5.17). Generally three phalanx marks (two interphlangeal marks and one basal phalanx mark) are observed in each finger except thumb, where two phalanx marks (one interphlangeal mark and one basal phalanx mark) are seen. All these phalanx marks are approximately equidistant from each other. But in reported case, prints of three members of a family including an identical twin pair and a member from their offspring show intraphalangeal mark, which is observed to be present transversely between the phalangeal marks (Figs. 5.12). This line was completely as like as that of phalanx mark (Figs. 1 and 5). From this observation of intraphalangeal marks among the members of the reported family and in niece of the twin pair, this trait is believed to be an inherited trait. Such variation in fingerprints is usually not seen in normal fingerprints of an individual, which is very unique.

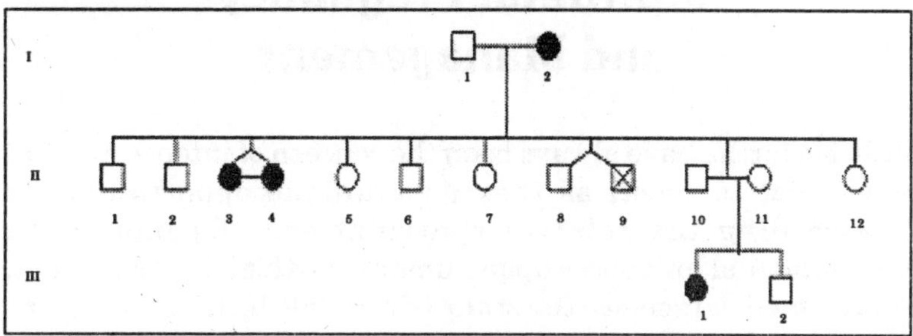

Fig. 5.17. Shaded circles indicate inheritance of a trait (intraphalangeal mark) among the members of a family.

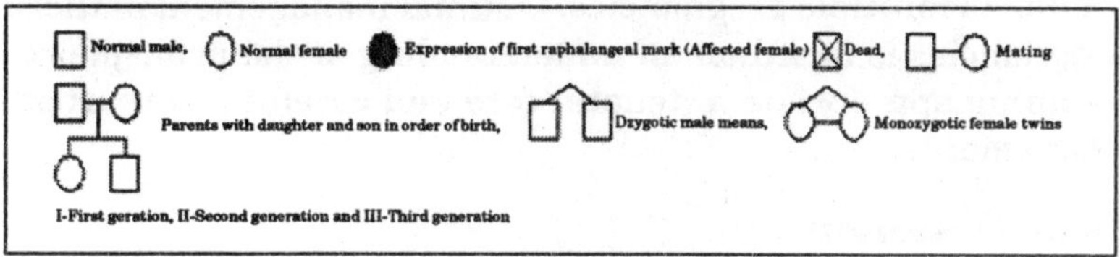

Basing on different types of pattern, intraphalangeal marks and phalanx marks, analyses are made on fingerprints of the members of the family and results are found out (Tables 5.1-4). The intraphalangeal marks of the studied case (Figs. 5.9-12) indicate the nature of inheritance of this trait. The presence of similar intraphalangeal marks in respective hands of the members of the identical twin pair indicates the genetic identicality. This is another clue to establish the identicality of twins. Further, multiple similar fingerprints from a crime spot may enable the forensic scientist to detect the criminals of a family or their relations having similar intraphlangeal marks.

6
Multifetal Pregnancy and Management

Multiple births have always been the concern of intense medical and social interest, as they do a unique opportunity for determining the relative contributions of genetic and environmental influences upon human growth and development. For the medical science the study of multiple births can present with a number of potentially difficult management decisions as they are frequently complicated not only by familiar problems such as hypertension and preterm labour but also by conditions unique to multiple pregnancies. Optimal management of these pregnancies is based on an understanding of the principles of twinning,appropriate antenatal care and careful intrapartum assessment.

Twin Gestation

Twin pregnancies have been an interest of the common and obstetric communities for centuries. After years of investigation and modifications in the management programmes, they still represent a high-risk pregnancy condition. At delivery, twins represent about 1% of births and 2% of infants, but they contribute 12% to the united states perinatal mortality rate of about 10 per 1000 births. Overall, the fetal mortality rate for twins is increased fourfold, the neonatal mortality rate is increased sixfold,and the perinatal mortality rate is increased tenfold over the rates for singletons. On a weight basis, the overall mortality rate for twins is increased by a factor of 1.5. For infants weighing less than 2.5 kg, twins have a better survival rate than singletons. For infants weighing more than 2.5kg, singletons fare better (Spellacy, 1994). Newman *et. al.*, (1989) have reported the median gestational age of triplets is substantially lower than twins at 31 to 35 weeeks. This median gestational age drops to 30.2 weeks in quadruplets (Elliot and Radin, 1992).

Etiology and Epidemiology

The overall incidence of twins at delivery is approximately one per cent. Depending on the release of ova, there are two types of twins, such as monoval and bioval. Monozygotic twinning is a chance occurrence and is little affected by other parameters. It occurs in approximately 3 to 4 of 1000 births throughout the world. Rates increase slightly with the delayed implantation, as occurs with in vitro fertilization.The frequency of dizygotic twinning does vary throughout the world and has several factors affecting its incidence. Heredity is one important factor. If the mother's first-degree relatives have had twins, the chance of her carrying twins is increased. If mother has already had one set of dizygotic twins, her chance of repeating in a subsequent pregnancy is increased twofold. There is no hereditary effect contributed by the father's side of the family.

Maternal race can affect the frequency of twinning. Monozygotic twinning frequency is unaffected by race, but dizygotic twinning occurs in approximately 7 to 10 per 1000 births for Caucasians, 10 to 40 per 1000 births for persons of African descent, and 3 per 1000 births for Asians. Maternal age and size also influence the rate of twinning with the highest frequency of dizygotic twinning occuring in mothers more than 35 years of age and in women who are obese.

The use of certain drugs in the preovulatory phase of the cycle increases the frequency of dizygotic twinning. These drugs include opiates and other compounds,responsible for induced ovulation. For example, the frequency of twinning with the use of Comid or gonadotropin releasing hormone (GnRH) is 10%, but with pergonal –induced cycles, it is 30% to 50%. Alterations in gonadotropin levels can produce multiple ovulations. Women who have spontaneously had dizygotic twins have been studied for their cyclic gonadotropin levels, but the results were inconsistent. In a few studies, preovulatory follicle-stimulating hormone levels have been found to be elevated. Because inhibin levels were normal, it was suggested that the GnRH may be increased in the cycles in which spontaneous multiple ovulations occur. However,many other studies have failed to find gonadotropin level alterations in women who previously had twin pregnancies. Few studies have been done of the cycle of the twin fertilization.

Diagnosis

Twins are diagnosed by clinical, biochemical, and biophysical methods. The clinical examination reveals a uterine fundal height growing at a rate that is greater than expected for a singleton pregnancy at that time. Later in gestation,two fetuses may palpated, and two different heart rates may be detected. Using only these clinical parameters, approximately one third of all twin pregnancies are missed and diagnosed only at delivery. The biochemical tests may demonstrate higher levels of fetal or placental substances in the maternal circulation than found during a singleton pregnancy. Higher levels of fetal or placental substances in the maternal circulation are found during a singleton pregnancy. Higher levels of maternal serum alphafetoprotein, human chorionic gonadotropin (hCG), human placental lactogen, and estriol suggest a twin gestation.

The definitive diagnosis is made with a biophysical test. In the past, X-ray diagnoses confirmed a twin gestation. Modern ultrasound studies can outline the twin pregnancy, its placentation,the structure of the central membrane and the fetal sexes. Errors in ultrasound diagnosis are most common when more than two fetuses are present. When a patient is sent for confirmation of a twin pregnancy and the ultrasonographer scans and finds two fetuses, the examination may then stop although more infants may be present. All patients with a twin pregnancy should have a thorough uterine examination to rule out possibility of three or more fetuses.

Maternal Complications with Twin Pregnancies

The physiologic changes occurring in the mother with a twin gestation are an exaggeration of what occurs in a singleton pregnancy. She has a greater increase in blood volume, pulse, cardiac output and weight gain. The average weight gain for a normal sized women with twins is 18.6 to 20 kg (41 to 44 lb) for the gestation. The following problems suchas, (*i*) preterm labor, (*ii*) hypertension, (*iii*) abruption, (*iv*) anemia (*v*) hydramnios, (*vi*) urinary tract infection, and (*vii*) postpartum hemorrhage and other maternal complications are significantly increased in women carrying a twin gestation.

(*i*) **Preterm Labor:** Two large studies of twin pregnancies demonstrated that the mean delivery time for twins was 37 weeks of gestation (256 and 258 days). Preterm labor for twin gestations is increased seven to ten times above singleton rates. It is a significant contributor to perinatal morbidity and mortality (Spellacy, 1994).

(*ii*) **Hypertension:** Women with a twin pregnancy have a twofold to fivefold increased frequency of hypertension complicating their pregnancy. This includes pregnancy-induced hypertension and preeclampsia with the accompanying proteinuria.

(*iii*) **Abruption:** Twin pregnancies have a threefold increased frequency of abruption, even when controlling for maternal hypertension. Although abruption occurs most frequently in the third trimester, it is also a significant risk immediately after vaginal delivery of the first infant,because the uterus changes its shape and sometimes shears off the placenta.

(*iv*) **Anemia:** The expanded maternal blood volume plus the increased demand by the fetus for two essential nutrients,folic acid and iron, contribute to the increased frequency of anemia. Women with twin pregnancies have two to three times the risk of developing anemia (haemoglobin concentration is less than 10 mg/ day) during their pregnancy. These women should receive supplements of folic acid (1 mg/day) and elemental iron (60 mg/day) during pregnancy.

(*v*) **Hydramnios:** Hydramnios occurs in 2% to 5% of twin gestations, and twin pregnancies account for approximately 8% to 10% of all cases of hydramnios. Acute hydramnios may accompany these pregnancies, particularly if there is a twin–to-twin transfusion with subsequent growth discordance. Medical management of these cases of hydramnios may be successful with drugs like indomethacin (100-150 mg/day). In the acute cases, aggressive removal of the fluid with multiple amniocentesis may be necessary.

(*vi*) **Urinary Tract Infection:** Women with multiple gestations have a 1.4 fold increased risk of developing urinary tract infection during pregnancy. This usually is a lower tract infection,because the incidence of pyelonephritis is not significantly increased.Preventive management involves urine cultures performed routinely during the gestation,one at the first prenatal encounter and the second at approximately 32 weeks of gestation. Any time when the woman is symptomatic for a urinary tract infection, an additional urine culture is necessary.

(*vii*) **Postpartum Hemorrhage and Other Maternal Complications:** The overdistention of the uterus frequently causes it to not contract well in the postpartum period. Approximately 10% of women delivering twins have postpartum uterine atony. Because postpartum bleeding is controlled by uterine muscle contraction, the atony may contribute to postpartum hemorrhage. Approximately 5% of women delivering twins have acute postpartum hemorrhage, which affects the management of the delivery. Blood may be made available before delivery, especially if the patient is anemic, and aggressive use of postpartum oxytocic drugs and uterine massage can be used to minimize bleeding. Other maternal complications have been suggested but are not routinely found. Placenta previa is occasionally listed as a problem,but in most large series, it has not been significantly increased in twin pregnancies. The glucose tolerance test studies of women with twins show a lower maternal glucose level than as seen in women with singletons, and no more glucose screening is necessary for twins pregnancies than for singleton pregnancies. Because of the frequent nonvertex presentation of one or more infants during labor, the use of cesarean section for delivery is increased.

Infant Complications with Twin Pregnancies

The following infant complications such as :

(*i*) prematurity,

(*ii*) congenital anomalies,

(*iii*) vascular communications between fetuses

(*iv*) discordance,

(*v*) vanishing twin,

(*vi*) monoamniotic twin pregnancy with cord enlargement,

(*vii*) dead fetus syndrome,

(*viii*) locking twins,

(*ix*) combined pregnancy,

(*x*) mole and fetus and

(*xi*) delayed delivery occur in twin pregnancies and they all increase the risk of perinatal mortality.

(*i*) **Prematurity:** As the average time to delivery of twin gestations is 37 weeks,the frequency of preterm delivery is significantly increased. This is a major contributor to neonatal morbidity and mortality.Most animal species have a maturation of their pulmonary function at approximately 85% of their average gestational age. Singleton infants have mature lungs at approximately 34 to 35 weeks of gestation,but twins reach this stage at 31 to 32 weeks of gestation.As a consequence, small infants weighing less than 2.5kg from a twin pregnancy are better in the nursery and have lower mortality rates than comparable weight infants from singleton pregnancies.

(*ii*) **Congenital Anomalies:** The frequency of congenital anomalies is increased two to three times in twins and this increase is mostly confined to monozygotic infants. Because the anomalies are often accompanied by a two-vessel umbilical cord,they can be screened for during the antepartum period with the routined ultrasound examinations.

(*iii*) **Vascular Communications between Fetuses:** With rare exceptions,vascular communications between twins are present only in monochorial placentas (Baldwins, 1991; Robertson and Neer, 1983). Nearly 100 per cent of monochorial placentas have vascular anastomoses, but there is marked variation in the number, size and direction of these seemingly haphazardly formed connections. Artery-to-artery anastomoses on the chorionic surface of the placenta have been identified in up to 75 per cent of monochorial placentas and are the commonest pattern. Vein-to-vein and artery-to-vein communica-tions are each found in approximately 50 per cent of similar placentas. One vessel may have several connections, sometimes to both arteries and veins. In contrast to these vascular connections on the surface of the chorion, there are artery-to-vein communications through the capillary bed of the villous tissue of the placenta. These deep arteriovenous anastomoses create a common villous district or third circulation that has been identified in approximately half of monochorial placentas.

Most of these vascular communications are hemodynamically balanced and of little fetal consequence. Rarely, however, they can cause hemodynamically significant shunts between fetuses. There are two patterns of hemodynamically significant anastomotic circulations namely acardiac twinning and the twin-to-twin transfusion syndrome. The incidence of the latter syndrome is unclear, but up to approximately a fourth of monochorial twins have some clinical features of this syndrome(Galea and Co-workers,1982).

(*iv*) **Discordance:** Discordance is defined as a significant difference in the weight of the two infants,variously set at more than 20% or 25% by different investigators,and the condition occurs in about 10% of twin pregnancies.

Two major contributors are known for discordance such

as, (*i*) the first is a significant difference in placental surface area for the two infants. The infant with the smaller placenta experiences growth retardation and near the end of gestation,has a high likelihood of demonstrating placental insufficiency and (*ii*) the second common cause for discordance is twin –to-twin transfusion. Blood is transfused from the donor twin to its recipient sibling such that the donor becomes anemic and its growth may be restricted, while the recipient becomes polycythemic and may develop circulatory overload manifest as hydrops. The donor twin is observed to be pale and its recipient sibling is plethoric. Similarly,one portion of the placenta often appears quite pale compared with the rest of the placenta.

The neonatal period may be complicated by dangerous circulatory overload with heart failure, if severe hypervolemia and hyperviscosity are not identified promptly and treated. Occlusive thrombosis is also much more likely to develop in this setting. Polycythemia may lead during the neonatal period to severe hyperbilirubinemia and kernicterus (Mahone *et al.,* 1993).

Both infants are at significant risk due to this problem, and it represents an extremely high risk perinatal event.The diagnosis at birth can be confirmed by a discrepancy between cord blood hemoglobin levels of greater than 5 mg/dL. During the antepartum period,when discordance of growth is suspected, ultrasound examinations can be performed. A difference in biparietal diameter of more than 5mm or a difference of estimated fetal weight by ultrasound examination of more than 25% suggests discordance. Confirmation is suggested if Dopler studies of umbilical artery flow show large differences. Twin-to-twin transfusions can be excluded if a diagnosis of dichorionic placenta is made. This can be determined using ultrasound and the rule of finding two infants of different genders; two separate placentas;or a central membrane that is more than 2 mm thick. Each of these methods has been used in the antepartum diagnosis to judge the potential risk. If acute polyhydramnios occurs with the twin-to-twin transfusin, the prognosis for fetal survival is poor, but the best management approach for this problem is aggressive mechanical reduction of the hydramnios by repeated amniocentesis.

(*v*) **Vanishing Twin:** Twins and higher order multiple pregnancies are more often conceived than born. During the first trimester of pregnancy, arrest of development and subsequent reabsorption of one or more of the fetuses may occur. This can be seen ultrasonographically and is known as the 'vanishing twin' phenomenon (Landy *et al.,* 1986). Vanishing bleeding in the first trimester may be related to the vanishing twin syndrome,but the prognosis for the remaining fetus after loss of a co-twin at this early stage of pregnancy is reported as good (Landy *et al.,* 1986).

Spellacy (1994) has narrated the vanishing twin syndrome refers to seeing twins on an early ultrasound examination but delivering only a singleton infant. With routine ultrasound studies in early pregnancy,it appears that the twin pregnancy frequency is close to 2%. Because only 1% of deliveries are twins, approximately one half of all twin pregnancies vanish. Most of the early gestations that are lost show a gestational sac without a fetal pole on ultrasound scans. An episode of vaginal bleeding usually accompanies the pregnancy loss. Serial studies of hCG levels show that they are lower in vanishing twin gestations than in normal twin pregnancies. After delivery of the surviving infant, the placenta frequently shows a whitish plaque on the membranes, which is the remnant of the other gestational sac. No harm is apparent for the mother or surviving infant.

(*vi*) **Monoamniotic Twin Pregnancy with Cord Entanglement:** If the monozygotic conceptus divides between days 8 and 13 of gestation, no central membrane is formed. The risk in these pregnancies is cord entanglement with fetal movement,which can lead to obstruction of blood flow and fetal death. Most of the fetuses that die are less than 32 weeks of gestation. The diagnosis can be suspected by being unable to find a central membrane on the ultrasound scan. If the scan shows a central membrane, ruling out this problem,the suspected diagnosis could be confirmed with an amniogram.After a gestation has passed 32 weeks, the risks for the infants is minimal, because their movement is limited and further entanglement is unlikely, so management can be expectant.

(vii) **Dead Fetus Syndrome:** Approximately 2% to 7% of twin pregnancies result in the death of one of the fetuses during the second half of gestation. If this occurs,it is extremely unlikely that maternal problems result, because rarely has significant hypofibrinogenemia been documented. Maternal fibrinogen levels decrease, but they usually plateau at about 150 mg/ml. Although some investigators have given mothers heparin as prophylaxis, others have achieved equally good results with no treatment.

There is a potential problem for the surviving infant. If the placenta is a dichorionic type, no vascular anastomosis exists, and the surviving fetus should fare well. However, with a monochorionic placenta, there is a high likelihood of vascular communication between the two twins. As soon as the first twin dies, there are vascular products that flow into the live twin. There may produce acute disseminated intravascular coagulation with multiple organ damage, particularly in the renal and central nervous system areas. Renal cortical cysts and multiple cystic lesions in the brain, known as multicystic encephalomalacia, are common. Approximately 25% of these infants die in utero, and as many as 50% of the survivors have brain damage. Because the process is rapid,the management needs to go for immediate delivery. If the surviving infant reaches a viable gestational age, even with rapid delivery,some survivors die (Spellacy, 1994).

(*viii*) **Locking Twins:** If twins are delivered vaginally and the first infant is breech with the second infant vertex,the two twins, heads can be interlocked with a chin-to-chin approximation. This is an infrequent complication with an incidence of about 1 in 90,000 pregnancies. It can be managed by turning the heads. It has been suggested that the infants can be replaced into the uterus, followed by a cesarean section delivery, although few cases have been reported.

(*ix*) **Combined Pregnancy:** The problem of a combined pregnancy occurs in approximately 1 of every 30,000 pregnancies. One conceptus is in the uterus, and the other is in an ectopic position. Operatve management of the ectopic pregnancy is needed early in pregnancy.

(*ix*) **Mole and Fetus:** Rarely, the twin pregnancy may have one placenta developing into a hydatidiform mole and the other conceptus developing as abnormal fetus and placenta.Ultrasound studies can diagnose this type of problem.

(*xi*) **Delayed Delivery:** Long intervals between infant deliveries have been reported for some pregnancies. This is important to record that if the first delivery occurs at a previable time,delays of 21 to 143 days have been reported, and such a delay could be long enough for the second twin to reach a viable stage of maturation.

MANAGEMENT

Antepartum Period

During the antepartum period, the emphasis should be on the prevention of recognized potential problems.The major problem of preterm labor can often be predicted by regular vaginal examinations using the cervical score. The cervical score represents the centimeters of thickness of the cervix (effacement) minus the centimeters of dilation. Little risk of premature delivery during in the next 7 to 14 days exists,until the cervical score is more than 0. Most controlled studies on the use of bed rest during the pregnancy to prevent preterm labor in a twin gestation show no significant effect. However, many of these were started late in the pregnancy, when rest was begun at 32 to 34 weeks of gestation. A few studies suggested some benefit in terms of lengthening gestation and increasing the weight of infants if the bed rest was introduced from week 24 of gestation and maintained through week 34 of gestation. Although hospital bed rest does not seem justified in terms of the cost, the use of home rest during this period does seem reasonable. It adds little cost to the pregnancy management, other than the loss of the mother's income. The patient could have some limited activities, such as bathroom privileges and sit-up eating, which improves compliance. Prophylactic tocolytics have been of no proven value in twin pregnancies. Although home uterine monitoring to detect uterine activity has proven unsuccessful in singleton pregnancies, the approach may be helpful in twin gestation.

If premature labour begins, the patient needs hospitalization with acute parenternal tocolytic management. Beta-mimetic tocolytics have a higher risk for pulmonary edema complications in twins than in singleton gestations, particularly if prolonged for more than 24 hours. If there is a question about maturation of the fetus, pulmonary maturation can be assessed with amniotic fluid, studies like the lecithin-sphingomyelin (L/S) ratio. The results are identical in the two twins if they are not discordant. If discordance exists, both sacs need to be tapped.

Twin gestations, particularly they are if not recognized, are at greater risk for folic acid deficiency. Apart from megaloblastic anemia, which occurs in up to 50 per cent of twin pregnancies in some series, folic acid deficiency has also been linked to abruption of placentae, pregnancy-induced hypertension, premature labour and stillbirth (Rothman, 1970). Because these women require higher concentrations of folic acid and iron,they need daily supplements of 1mg of folic acid per day and 60 mg of elemental iron. There is no evidence that other prenatal vitamins are of any help.

Because twin discordance results from twin-to-twin transfusions, establishing the type of placentation early in pregnancy is important. If a dichorionic placenta is established by ultrasound rule-of -two, surveillance for this problem is not necessary. If there is a monochorionic placentation, serial ultrasound assessments for discordance are needed.

Prenatal genetic testing is important because of the increased frequency of congenital anomalies among twins. It appears that the risk for chromosome abnormalities is higher in twins gestations, and genetic analysis by means of amniocentesis or chorionic villus sampling should be recommended for a twin gestation if the mother is 33 years of age, compared with 35 years of age for singleton. Maternal serum alpha-fetoprotein levels are elevated in twin gestations and the median level for twins at 16 weeks of gestation is 2.5 times the multiple of the median, compared with two times for singletons. If genetic analysis using aminocentesis is required, both sacs should be tapped. Using a glucometer and glucose as a marker, the examiner can identify quickly that independent sacs have been tapped. The use of colored dye as a marker, the examiner can identify quickly that independent sacs have been tapped. The use of colored dye as a marker should be avoided. Methylene blue for example has been implicated in causing fetal hematolytic anemia and jejunal atresia.

If hypertension or other problems that increase the risk for the fetus occur, serial fetal surveillance should be performed. This can be adequately done using fetal heart rate monitoring and doing weekly non-stress tests on each infant.A non-reactive test needs follow up with a contraction stress test. If preterm labor has already threatened the pregnancy, a biophysical profile could be substituted. Good perinatal mortality data have resulted from such testing. Regular ultrasound examinations should also be done to verify adequate growth rates for each infant. The ultrasound examination should document fetal growth,amniotic fluid volume, the central membrane, and anomalies. Because the average delivery time for twins is 37 weeks of gestation, the physician may assume a twin gestation is in the postdate period after it has reached 39 or more weeks of gestation. Appropriate surveillance testing should be started at that time.

Intrapartum Period

The mode of delivery is often influenced by the presentations of the twins(Grant, 1989)which may be divided into three groups such as (*i*) first twin vertex second twin vertex, (*ii*) first twin vertex second twin non-vertex and (*iii*) first twin non-vertex.

Studies performed Spellacy (1994) during early labor show that approximately one half of the twin infants will be vertex presentations and that the other one half will be breech or transverse lie presentations. If both infants are vertex,most obstetricians allow normal labor and deliver of each infant. Before electronic fetal heart rate monitoring began in the 1950s, it was shown that the optimal time for vaginal delivery between the two infants was 5 to 15 minutes. With continuous heart rate monitoring of the second infant after the first is delivered, there appears to be no optimal time for delivery and no operative intervention is necessary if labor is progressing. During labor, both fetuses should be monitored with electronic continuous heart rate recordings. Early cord clamping is important after the first delivery, because excessive bleeding of the cord could exsanguinate the second twin if a monochorionic placental vascular anastomosis existed. The risk abruption increases with delivery of the first infant and this situation must be monitored carefully. If vaginal delivery is attempted, the delivery room should be set up for a possible emergency cesarean section of the second infant. An ultrasound machine should be available to determine the presentation of the second infant immediately after delivery of the first. Support staff, including an anesthesiologist and neonatologist should be present. Cord hematocrits should be obtained if same-sex twins are delivered, and these bloods should be sent to the laboratory immediately because they may disclose a twin-to-twin transfusion problem.

If the first twin presents vertex and the second presents breech,many obstetricians deliver the first infant vaginally and attempt an external version on the second infant to allow it to deliver vaginally with a vertex presentation. Because obstetricians often deliver breeches by cesarean section,many twin presentations in which the second infants is not in a vertex position in early labor, undergo elective ceasarean section for both. The overall cesarean section rate for twins approximate 50%.

Postpartum Period

Because of the potential risk for uterine atony and postpartum hemorrhage, the mothers should be closely monitored during the first three hours after delivery of twins. Adequate oxytocics should be administered and the uterine fundus should be regularly massaged to be certain that good uterine tone exists.An intravenous access line must be in place and if the mother began labor with anemia, blood products should be available in case of significant acute hemorrhage. The maternal task of caring for twin infants is often overwhelming. Dialogue with and support for the mother in the early weeks after delivery are important. Postpartum depression is more common in women delivering twins (Spellacy, 1994).

7
Multiple Birth

Multiple births may occur by subdivision of one zygote into more than two parts, by the simultaneous fertilization of more than two ova, or by a combination of both these factors (Figs. 7.1 and 7.2). Humans have always been fascinated by multiple births. Multiple pregnancy rates vary world wide. Multiple pregnancies have always been the subject of intense medical and social interest, as they offer a unique opportunity for determining the relative contributions of genetic and environmental influence upon human growth, development, behaviour and intelligence.

Study on five sets of triplets (Figs. 7.3 KHU 1- 7 NGH 1), morphological, physiological and dactylograhic parameters show less percentage of concordance in the members of each set of triplet than members of identical twins except one set of triplet, where out of three members, two members show more similarity on the different traits. Investigation shows multiple births include both identical and non-identical individuals or all non-identical individuals. The spontaneous occurrence of triplets can be predicted by the Hellin rule: if the frequency in a population of twins is n, then that for triplets is n^2. In the 1990s, when assisted reproductive technologies are commonly used for induction of ovulations and pregnancies with two to eight fetuses were not uncommon. Although the spontaneous frequency of triplets is approximately 1 in 10,000 pregnancies, the actual frequency is much higher.

Little is known is about triplet pregnancies, because few large series have been described. The usual maternal weight gain is approximately 20. 5 to 23 kg (45-50 lb). Spellacy (1994) reports the usual spontaneous time for delivery is 32 to 34 weeks,

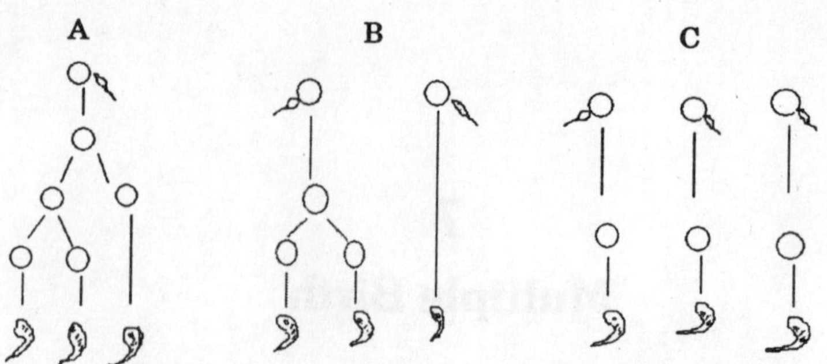

Fig. 7.1. Derivation of triplets (A) from one ovum, (B) from two ova and (C) from three ova.

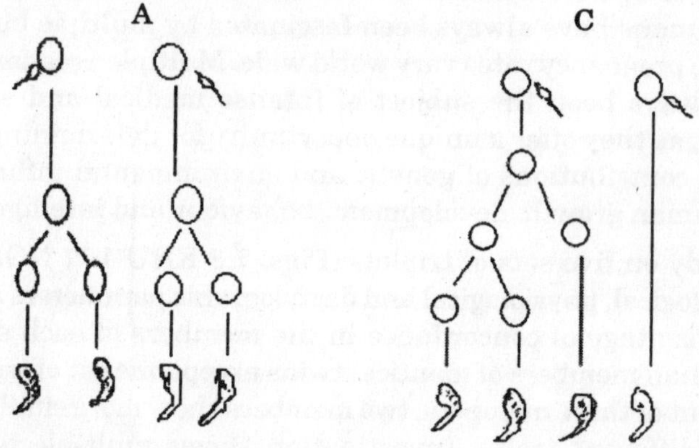

Fig. 7.2. Two ways of formation of quadruplets from two ova.

Fig. 7.3.

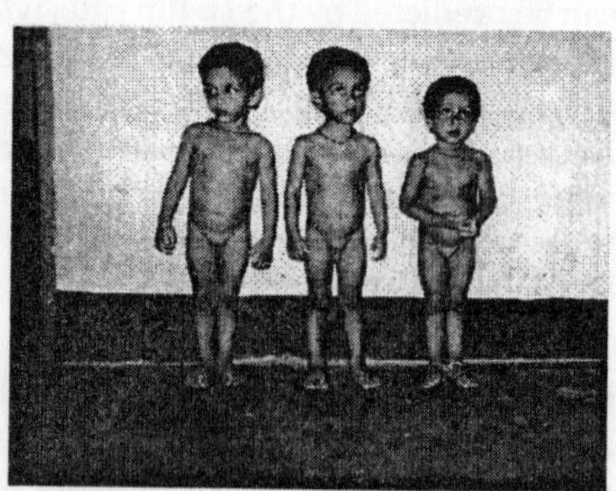

Fig. 7.4.

Fig. 7.5.

Fig. 7.6. **Fig. 7.7.**

Fig. 7.7 . Triplets (in Fig. 7.5, out of three individuals one was dead)

and the average infant weight is 1.8 to 1.9 kg. As the number of fetus increases, the duration of gestation decreases. Approximately half of twins deliver at 36 weeks or less. The mean age at delivery of triplets is 33.5 weeks, with 90 per cent, 24 per cent and 8 per cent delivered before 37, 32 and 28 weeks respectively (Berkowitz *et. al.*, 1996). Fetal growth is normal in the first two trimesters but begins to slow down compared with singletons by 27 weeks of gestation. For twins, the slow down occurs at about 30 weeks of gestation.

Discordance among triplets is more common with approximately 30% of these pregnancies having a discordance of weight at least 25%. In twin gestation, approximately 10% have that degree of weight discordance. Studies have shown that an early pregnancy reduction of fetuses from three to two does not increase survival rates, and no attempt is recommended. The perinatal mortality rate for triplets is similar to that for twins. The maternal complications and infant complications are similar to those for twins. Most obstetricians deliver triplets by ceasarean section because of the likelihood that one or more of the infants are in a nonvertex presentation.

When a woman has four or more infants in utero, the overdistention of the uterus leads to much earlier preterm labor. Most of these pregnancies are the result of induced ovulation or in vitro fertilization. Spellacy (1994) has reported for quadruplets, the average maternal weight gain is about 23 to 25 kg (50-55 lb) and the average time for delivery is 30 to 31 weeks. The average infant weight is approximately 1. 2 to 1. 5kg. The average age at delivery was 31 weeks in 37 quadruplet pregnancies (Barton and colleagues, 1996). A reduction in fetal numbers during the first trimester has been successful in bringing the number of infants to two or three and improving the perinatal survival rate.

Although quintuplets or more are born successfully, the rate of survival of all the infants is very remote.

In this study of multiple births the pedigree analyses reveal the birth of twins in both maternal and paternal generations of twins and of triplets. Thus, it indicates the contributions not only of the female but also of the male for twinning and multiple birth tendency. Most of the eutherian and metatherian mammals are polyovulatory as a result of which multiple births occur. Some species (*e.g.*, marmosets) are diovulatory and discharge two ova in each ovarian cycle resulting in twins. Thus, this twin study about the presence of twins both in maternal and paternal generations of the twins and multiple births indicates twinning and multiple births are accidental and sporadic in man and are influenced by genetic and environmental factors though they are monoovulatory in the nature of ovulation.

Further, the extent to which identical and non-identical differ in their degree of concordance offers a measure to assess the relative roles of environment and heredity. Thus, more concordance in morphological and physiological parameters and dactylography of identical twins signifies a great influence of genetic similarity on the concordant appearance as such traits have strong genetic component. This study also implies the majority of the genes of identical twins are indeed identical, but some changes are possible owing to replication errors and other types of mutations as a result of which it might be resulted into percentage of discordant in identical twins.

The frequency of monozygotic twin birth is relatively constant worldwide, at approximately one set per 250 births and is largely independent of race, heredity, age and parity. The frequency was once thought to be independent of therapy for infertility; however, it is now observed that the incidence of zygotic splitting is increased following assisted reproductive technologies (ART). The advent of ovulation induction (OI), ovarian hyperstimulation and assisted reproductive techniques (ART) [*in vitro* fertilization, gamete intrafallopian transfer(GIFT), sperm intra-fallopian transfer (SIFT), and zygote intra-fallopian transfer (ZIFT)] have in recent years had a large impact on the incidence of multiple gestation, particularly gestations of high order. As many as 80 per cent of triplets and nearly all high-order multiple gestations are initiated through such techniques. However, the incidence of dizygotic twinning is influenced remarkably by race, heredity, maternal age, parity and especially fertility drugs. The frequency of multiple fetal births varies significantly among different races and ethnic groups. Myrianthopoulos (1970) identifies the birth of twins in 1 of every 100 pregnancies among white women, compared with 1 of 80 pregnancies for black women. In some areas of Africa the frequency of twinning is very high whereas twinning in Asia is less common.

As a determinant of twinning, the family history of the mother is more important than that of the father. The marked difference in twinning frequency may be the consequence of follicle stimulating hormone (FSH) level, which can lead to multiple ovulation. The rate of twinning rises from zero at puberty, a time of minimal ovarian activity, to a peak at age 37, when maximal hormonal stimulation increases the rate of double ovulation. The fall in incidence after age 37 probably reflects exhaustion of the Graafian follicles. Increasing maternal age and parity have been shown to increase the incidence of twinning. The common factor linking race, age, weight and fertility to multiple gestation may be follicle stimulating hormone levels. The induction of ovulation by use of gonadotropins (follicle stimulating hormone plus chorionic gonadotropin) or clomiphene remarkably enhances the likelihood of multiple ovulations. It is also observed that techniques designed to increase the probability of pregnancy increase the probability of multiple gestation as well.

Determination of zygosity is another important factor of twin study since it facilitates inter-twin organ transplantation later in life. Although it is observed that identical twins look alike and non-identical twins look dissimilar, but in rare cases it may not be the situation. Monozygotic twins may actually be discordant for genetic mutations as the result of a postzygotic mutation, or may have the same genetic disease but with marked variability in expression. Most interestingly, monozygotic twins may be discordant for malformations, which involve asymmetrical organs. For example, a fetus who is the mirror image of his twin may have a heart defect caused by reversed looping or laterality. However, for genetical, embryological and obstetrical reasons, it is more important to determine the number of chorions. In fact, chorionicity can be determined by as early as the first trimester using multiple sonographic signs.

An examination of the placenta and membranes serves to establish zygosity promptly in about two thirds of cases. Delivery of the placenta should be accomplished with care to preserve the attachment of the amnion and chorion to the placenta, because identification of the relationship of the membranes to each other is critical. With one common amniotic sac, or with juxtaposed amnions not separated by chorion arising between the fetuses, the infants are monzygotic. If adjacent amnions are separated by chorion, the fetuses could be either dizygotic or monozygotic, but dizygosity is more common. Of course, different blood types confirms dizygosity, although demonstrating the same blood type in each fetus is not enough to confirm monozygosity. For definitive diagnosis, an accurate technique such as DNA fingerprinting can be suggested.

Twins of opposite sex are almost always dizygotic. Very rarely, monozygotic twins may be discordant for phenotypic sex. This occurs when one twin is phenotypically female due to Turner syndrome (45, X) and its sibling is 46, XY. A maternal family history of twins, older maternal age, high parity, large maternal size and a previous history of twins in believed to provide some clues, but knowledge of recent administration of either clomiphene or gonadotropins or pregnancy accomplished by assisted reproductive technology would provide strong ones. Although this study offers some light on the genetics of twins, much detailed dactylographic, biochemical, embryological and hormonal analyses are suggested for understanding the possible reasons of human twinning, frequency of twinning, zygosity, twin-to-twin transfusion syndrome (TTS) and etiology of multiple fetuses.

Glossary

abortion - The expulsion of a human fetus within the first 12 weeks of pregnancy, before it is viable. The arrested development of an embryo or an organ at a more or less early stage.

accidental pattern - Some composite patterns are called accidental as they rarely occur, such as loop-by-loop, whorl resting on loop, loop resting on whorl, whorl resting on whorl, arch with pocket. Loop by loop is a lateral pocket.

aetiology - (US etiology) [Gr. *aitia* a cause; *logos* study] -the causation of diseases and disorders as a subject of investigation.

allantois - One of the extra-embryonic membrane arising a pouch from hindgut in the embryos of amniotes(reptiles, birds and mammals).

amnion - (Gr. *amnion* - fetal membrane) the innermost membrane that encloses the embryo of a mammal, bird or reptile.

amniocentesis - (Gr. *amnion* - fetal membrane;*kentesis*-pricking) A process in which amniotic fluid is sampled using a hollow needle inserted into the uterus, to screen for abnormalities in the developing fetus.

amniotic fluid - The fluid surrounding a fetus within the amnion.

ancestor - One from whom a person is descended; fore father; progenitor. The actual or hypothetical form or stock from which an organism has developed or descended.

anemia - A disorder characterized by deficiency of haemoglobin content in blood due to reduced number of red blood cells.

antenatal – before birth; during or relating to pregnancy : antenatal care.

antepartum - adj. occurring not long before childbirth. ORIGIN mid 19th cent. from Latin, 'before birth'.

arch - A pattern in which the ridges extend from one side of the finger to the other. Without turning back, rising slightly at the center, where the curvature looks like an arch. There is ordinarily no delta.

central pocket loop - In central pocket loop pattern, there is a combination of loop and whorl. The whorl formation lies in the core region of the pattern in the form of a pocket, which is formed when one or more ridges immediately about the core deviate in course from the general course of the other recurring ridges.

cephalopagus - (Gr. *kephale* - head; *pagus*-combining form) In cephalopagus twin, the anterior (front of the body) upper half of the body gets fused with two faces on opposite sides of a conjoined head. The heart may also be conjoined or shared. Their survival is very difficult.

cephalothoracopagus-(Gr. *kephale* - head; *thorax* - chest; *pagus* - combining form). Fused from the head to chest. There is only one brain and a head. There may be a rudimentary second face. They usually share a heart and have fused gastro-intestinal tracts and are non-viable. They are similar to syncephalus, but the chest is fused.

chimera or chimaera -(Gr. *chmaira* - a she-goat) literally means wild fancy or a picture of an animal having its parts made up of various animals. The general biological explanation of chimera is an organism, which is made up of two genetically distinct tissues. A chimera may be defined as an organism whose cells originate from two or more zygotes.

chorion - (Gr. *chorion* - skin) The outermost membrane surrounding an embryo of a reptile, bird or mammal. In mammals (including humans) it contributes to the formation of the placenta.

composite - Compound, that is, not simple. The same meaning applies to this word when used in the science of fingerprints. All those impressions which show a combination of two or more ridge designs with two or more deltas are grouped under composites.

conjoined twins - Identical twins who develop with a single placenta from a single fertilized ovum and are always of the same sex. Also called as Siamese twins arise by incomplete separation of fertilized ovum at about 15 days or more after zygote formation. Such twins are almost equal in sizes and receive proper blood supply. In Siamese twins, the two partners may be completely separated by the appropriate surgery. Most of the conjoined twins do not survive due to complication at parturition.

core - The core means the central part of a pattern. The type of the core varies according to the pattern.

craniopagus - (L. L. *cranium* – the skull; *pagus* - combining form) Twins joined at the skull, but not at the face or base of the skull, and share bones of the skull and often have fused brain surfaces. They can be joined crown to crown (vertical), at the back of the head (occipital), front of the head (frontal) or the side of the head (parietal or temporal).

dactylography - (Gr. *daktylos* - finger; *graphein* - to write) The scientific study of fingerprints.

delta - The word 'delta' is the fourth letter of the Greek alphabet and corresponds to the English letter 'D'. This Greek letter is triangular in shape; hençe the word is applied to many things shaped like triangles such as island at the mouth of rivers. When a ridge bifurcates, and the two arms of the bifurcating ridge diverge causing an interspace within which the pattern lies, the point of bifurcation is called "delta".

dibrachius dipus(Gr. *dis* - twice; *brachion* - arm; *dis* - twice; *pous, podos* - foot) Twins are characterised by two arms, two legs, *i.e.*, "normal" body lay out.

dicephalic parapagus (Gr. *dis* – twice, *kephale* – head; *para* – beside; *pagus* - combining form) Twins characterised by one trunk with two heads two arms or four arms and two legs.

dicephalus – Conjoint twins of two distinct heads and single body.

diproscopic parapagus (Gr. dis – twice; *pro* –before; *skopeein* - to see; *para* – beside; *pagus* - combining form) Twins having a single head with two faces.

diproscopus – conjoint twins of two faces, one head and single body.

dithoracic parapagus (Gr. *dis* – twice; *thorax* – chest; *para* – beside; *pagus*-combining form) Twins joined at the abdomen and pelvis, but not the chest and characterised by the presence of two chests, two heads, four arms and two legs.

dizygotic twins - Non-identical twins arise from the simultaneous release of two separate eggs from the ovaries of the mother each fertilized by a separate sperm cell. Therefore, these twins are referred to as dizygotic twins (two eggs).

duplicata incompleta (L. *duplicare*, atum – *duo*, two; *plicare* – to fold; *in*-not; *completus* – filled, finished) Twins are observed as any part, which has not completely separated. They are dicephalus having two heads and one body.

fetus - (pl. fetuses) (L. fetus 'pregnancy, childbirth, offspring') an unborn or unhatched offspring of a mammal, in particular, an unborn human more than eight weeks after conception.

epidemiology - The branch of medicine which deals with the incidence, distribution and possible control of diseases and other factors relating to health.

fertilization - union of a mature ovum and a mature sperm to form a zygote.

gestation - period between fertilization and birth, i. e, period of pregnancy, in which young is carried in uterus before birth.

haemorrhage - (Gr. *haima* – blood; *rhegnynai* – to burst) A discharge of blood from the blood vessels.

ischiopagus (Gr. *ischion* – the hip joint; *pagus*-combining form) Joined from the navel to a large conjoined pelvis joined end to end with the spine in a straight line are the features of Ischiopagus twin. They have four arms, a variable number of legs, usually single external genitalia and a single anus

lateral pocket loop - A type of pattern, which consists of two inter clasped loops. One overlapping the other and forming a definite side pocket.

loop - A pattern in which one or more ridges recurve, that is, run back on their previous course, making a half turn or more around the core, having a delta and at least one ridge intervening between the inner and outer termini.

menopause - The period of permanent cessation of menstruation, usually occurring in women between the ages of 45 and 50.

Menstruation - The act of periodically discharging blood and mucosal tissue from the uterus, occurring approximately monthly from puberty to menopause. The period of menstruating.

monozygotic twins - Twins arising from a single fertilized ovum. These are called monozygotic or maternal twins. The genetic constitution of the two twins is exactly the same. Hence they are of the same sex. They are also exactly alike in appearance.

monochorionic biamniotic type - At the blastocyst stage, the inner cell mass is separated completely into two equal parts and each part develops into a separate embryo. This variety of twins presents single placenta, common chorionic sac, but each embryo is invested by a separate amniotic membrane.

monochorionic monoamniotic type - When the bilaminar germ disc is formed within the blastocyst cavity, sometimes two separate organizing centers appear in the disc (instead of one), and differentiate into separate embryos. The twin embryos lie within a common chorionic sac and are enveloped by a single amniotic membrane.

monozygotic bichorionic type - At two-cell stage of cleavage division, the totipotent blastomere cells develop into two separate zygotes within the common zona pellucida. When the zona pellucida disappears, the two zygotes are implanted separately in the uterine edometrium. Therefore, this variety of twins possesses separate placenta and separate chorionic sacs, like dizygotic twins.

mosaicism (Fr. *mosaique*, L. L. *masaicum*, Gr. *mousa* - a muse) the phenomenon of the presence of patches of tissues of unlike constitution side by side. A mosaic is simply a hybrid with the parental characters side by side, which is not blended. A mosaic consists of two different types of cells, originally the same at the stage of the zygote, but becoming genetically different subsequently by some event such as mutation, by nondisjunction in one cell line, or by X-chromosome inactivation.

omphalo-ischiopagus(Gr. *omphalos* - navel; *ischion* - the hip joint; *pagus*-combining form) It is the combination of omphalopagus and ischiopagus condition. They are joined as ischiopagus twins, but face to face with joined abdomen and fused pelvis. They have four arms and a variable numbers of legs (2 or 3 or 4).

omphalopagus (Gr. *omphalos* - navel; *pagus* - combining form) They are joined face to face at the navel and sometimes also the lower chest. They have two separate hearts, but the pelvis may be conjoined(two pelvis) having four arms and four legs.

parapagus (Gr. *para* - beside; *pagus* - combining form) Twins that share a conjoined pelvis fused side by side. In this case some parts of the pelvis may be duplicated. Parapagus twins may be dipus, having two legs, tripus, having three legs, tetrapus, having four legs. Parapagus twins may also be dibrachius having two arms, tribrachius having three arms or tetrabrachius having four arms.

parasitic twins –In symmetrical Siamese twins one member is fairly normal (the host or auto site) but the other member is a parasite, which is severely underdeveloped and dependent on its host for nutrition. Such parasites may often be removed by surgery. One partner of the conjoined twins is rudimentary due to diminished blood supply. Sometimes the parasitic twins may be completely enclosed within the body of the co-twin. This phenomenon is known as "fetus in fetus".

parturition - (L. *parturire*- to bring forth) The process of birth of mammals.

paternal – Characteristic of or befitting a father; fatherly : a kind and paternal reprimand. 2. of or pertaining to a father. 3. related on the father's side : one's paternal grand father. 4. derived or inherited from a father : paternal traits.

pattern - Whenever an interspace is left between the boundaries of a different system of ridges, it is fitted by a small system of its own, which will have some characteristic shape and is called a 'pattern'.

pedigree – An ancestral line; line of descent; lineage; ancestry. 2. a genealogical table, chart, list, or record : a family pedigree.

Placenta – A vascular structure, embryonic and maternal, through which embryo and fetus are nourished while in uterus.

post-partum – (L. *post* - after; *parere* - to bring forth) After parturition.

Puberty – The time of beginning of sexual maturity in an individual.

pygopagus - (Gr. *pyge* - rump; *pagus* - combining form) Twins are identified from the joining at the perineum and may share part of the spinal cord. They have a single anus, two rectums, four arms and legs.

rachipagus(Gr. *rachis* – spine; *pagus* combining form) Twins fused back to back above the sacrum, involving different segments of the spine.

radial loop - The ridges terminate in the direction of the radial bone of fore arm, that is the ridges slant towards left in case of right hand fingers and to the right in case of left hand fingers.

siamese twins - Identical twins but they are occasionally born physically in conjoint form. The original Siamese twins, Chang and Eng, born in 1811 of Chinese parents in Siam (Thailand) were joined by a ligament from breast bone to navel.

superfecundation - The phenomenon of production of members of same multiple births by different fathers is known as superfecundation and is well known in domestic animals.

twin (O. E. *twinn* – double) Born two at a birth : a type of multiple birth producing two offsprings at a time. It is of two types : (i) identical or monozygotic or uniovular and (ii) non-identical or dizygotic or biovular or fraternal twin. Identical twins are developed from a single zygote where as non-identical arise from two different zygotes formed by fertilization of two eggs by two sperms.

syncephalus - (Gr. *syn*-together; *kephale*-head; *pagus*-combining form) A form of cephalopagus twins joined by the face and has a single head and two bodies. Syncephalus twins usually have one head and a single face with four ears, two on the back of the head and the other two in the normal position.

syndelphus (Gr. *syn* – together; *adelphos*- brother)Twins having one body and eight limbs.

syndrome - (Gr. *syndrome*- symptoms)Occurrence of group of symptoms : a characteristic pattern.

thoraco-omphalo-ischiopagus - (Gr. *thorax* - chest; *omphalos* navel; *ischion* – the hip joint; pagus-combining form) This is just like omphalo-ischiopagus but also joined at the chest. They bear two heads, four arms and a variable numbers of legs (two or three or four).

thoraco-omphalopagus (Gr. *thorax* – chest; *omphalos* – navel; *pagus*-combining form) A combination of thoracopagus and omphalopagus state. They are conjoined from the upper chest to the navel and either share a single heart or have conjoined hearts. They also have shared livers and gastro-intestinal tracts, four arms and four legs.

thoracopagus (Gr. *thorax* - chest; *pagus* - combining form) Joined face to face from the upper chest down to the navel, with a shared heart of conjoined heart, four arms, four legs and two pelvises.

trophoblast – The outer layer of cells of a blastocyst that adhere to the endometrium during implantation.

twinned loop - Two well-defined loops, one loop, more or less, surrounds the other.

ulnar loop- The ridges terminate in the direction of ulnar bone of the fore arm, in other words, the ridges slant towards right in case of right hand fingers and towards left in case of left hand fingers.

whorl - The word 'whorl' in the English language means a number of leaves in a circle round the stem; but in relation to recurring ridges, it represents a type of pattern in which the extent of curvature around the core is one complete twist or circuit, and there are two deltas. There may be one or more ridges making a turn at least one complete circuit.

xipho-omphalo-ischiopagus - (Gr. *xiphos* – a sword; *omphalos* – navel; *ischion* – the hip joint; *pagus*-combining form) Similar to omphalo – ischiopagus but they are joined from the lower chest. They have two heads, four arms and a variable number of legs (two or three or four).

Xiphopagus - (Gr. *xiphos*- a sword;*pagus*-combining form) The individuals joined at the xiphoid process, i. e., roughly from the navel to the lower breastbone. They usually have separate vital organs except for a shared or conjoined liver.

References

Alexander, D. (1975) *Thumb Impression Identification Simplified "Faith Cottage"*. Allahabad, India, pp. 1-90.

Altenburg, E. (1957) *Genetics. Halt, Rinehart and Winston,* Inc., New York, USA, pp. 1-488.

Anonymous (2005) Twin brothers: twin number. *The Samaja*, 76(159): Sunday Special, pp. 06.

Atherly, A. G.; Girton, J. R. and Mc Donald, J. F. (1999) *The Science of Genetics*. Saunders College Publishing, United States of America, pp. 1-704.

Bajpai, S. R. (1960) *Methods of Social Survey and Research*. 15th Edition. Kitab Ghar, Kanpur, pp. 1-405.

Baker, J. Newton (1955) *Law of Disputed and Forged Documents.* The Michie Company, Charlottesville, Virginia, pp, 1-560.

Baldwin, V. J.; Pathology of multiple pregnancy. In Wigglesworth, J. S. and Singer, J. (eds)(1991) *Text book of fetal and perinatal pathology*. Boston, Blackwell, pp. 238.

Balinsky, B. I. (2004) *An Introduction to Embryology*. Fifth Edition. Thomson Asia Pte. Ltd., Singapur, pp, 1-715.

Barton, J. R.; O'Brien, J. M.; Jacques, D. L.;Bergauer, N. K.;Stanziano, G. J. and Sibai, B. M. (1996) Perinatal outcome in quadruplet gestations. Am. *J. Obstet. Gynecol.*, 174: 478.

Bates, Billy Prior (1970) *Identification System for Questioned Documents* (I. S. Q. D). Charles C. Thomas, Springfield, Illinois, USA, pp. 1-97.

Bedford, J. M. (1977) Sperm egg interaction-the specificity of human spermatozoa. *Anat. Rec.*, 188:477-488.

Bergsma, D. (1965) (Ed.) New Directions in Human Genetics. Williams and Wilkins Co., Baltimore for the National Foundation March of Dimes, *White Plains, New York*, 1 (2): 79.

Berkowitz, R. L.; Lynch, L.; Stone, J. and Alvarez, M. (1996) The current status of multifetal pregnancy reduction. *Am. J. Obstet. Gynecol.*, 174:1265.

Bodmer, W. F. and Cavalli-Sforza, L. L. (1970) *Intelligence* and race. *Sci. Am.*, 223: 19-29.

Bodyazhina, V. I. (1987) *Textbook of Obstetrices*. MIR Publishers, Moscow, pp. 1-399.

Bose, H. C. (1927) *Finger Print Companion*. Second Edition. Finger Print Bureau, Writers' Buildings, Calcutta, pp. 1-112.

Brewester, F. (1936) *Finger Prints*. The Eastern Law House, Calcutta, pp. 1-234.

Bridges, B. C.; O' Hara, C. E. and Vollmer, A. (1963) *Practical Finger Printing*. Funk and Wagnalls Company, Inc., New York, pp. 1-370.

Brown, T. A. (2002) *Genomes*. Second Edition. John Wiley and Sons, Inc., New York, USA, pp. 1-503.

Browne, D. G. and Brock, A. (1953) *Finger Prints. George* G. Harrap and Co. Ltd., London, pp. 1-218.

Bulmer, M. G. (1970) *The Biology of Twinning in Man*. Clavendon. Press, Oxford, pp. 1-232.

Carlson, B. M. (1994) *Human Embryology and Developmental Biology*. Mosby Year Book, Inc., St. Louis, pp. 41-44.

Chang, M. C. and Hunter, R. H. F. (1975) Capacitation of mammalian sperm. In: *Hand book of physiology,* Section 7, Vol. V, Washington D. C., pp. 339-352.

Chatterjee, C. C. (1988) *Human Physiology*. Vol. II. Medical Allied Agency, Old Ballygunge, Second Lane, Calcutta, pp. 11-15.

Chatterjee, C. C. (1992) *Human Physiology*. Vol. I. Medical Allied Agency, Mahatma Gandhi Road, Calcuta, pp. 1-726.

Clermont, Y. (1963) The cycle of the seminiferous epithelium in man. *Am. J. Anat.*, 112: 35-52.

Closinger, C. R.; Bohman, M. and Sigvarodsson, S. (1981) Inheritance of alcohol abuse: cross fostering analysis of adopted men. *Arch. Gen. Psychiatry*, 38: 861-868.

Cohen, J. (1977) *Reproduction*. Butterworths, London, Boston, Sydney, Wellington, Durban and Toronto, pp. 1-290.

Cole, H. H. and Cupps, P. T. (1964) *Reproduction in Domestic Animals*. Academic Press Inc., Ltd., London, pp. 1-589.

Crosignani, P. G. and Mishell, D. R. (1976) *Ovulation in the Human.* Academic Press Inc., London, pp. 1-311.

Dal Colletto, G. M.; Fulker, D. W.; Barretto, O. C. and Kolya, M. (1993) Genetic and environmental effects on blood cells. *Acta. Genet. Med. Gemellol.* (Roma), 42(3-4):245-252.

Datta, A. K. (2004) *Essentials of Human Embryology*. Fourth Edition. Current Books of International, Calcutta, pp 1-307.

David, M. E.; Frazer, I. H.; Boomsma, D. I. and Martin, N. G. (2001) Developmental genetics of red cell indices during puberty: a longitudinal twin study. *Int. J. Hum. Genet.*, 1(1): 41-53.

Dosso, A. A. and Bovet, J. (1992) Monozygotic twin brothers with age-related macular degeneration. *Ophthalmologica*, 205: 24-28.

Edwards, R. G. and Fowler, R. E. (1970) Human embryo in the laboratory. *Sci. Am.*, 223:44.

Elliot, J. P. and Radin, T. G. (1992) Quadruplet pregnancy : contemporary management and out come. *Obstet. Gynecol.*, 80:421-424.

Erlenmeyer-Kimling, L. and Jarvik, L. (1963) Genetics and intelligence: a review. *Science,* 142: 1477-1479.

Evans, D. M.; Frazer, I. H. and Martin, N. G. (1999) Genetic and environmental causes of variation in basal levels of blood cells. *Twin res.*, 2(4): 250-257.

Eysenck, H. J. (1990). Genetic and environmental contributions to individual differences: the three major dimensions of personality. *Journal of Personality*, 58: 245-261.

Field, J. F.; Masson, C.; Robins, V. and Bradley, J. (2002) (Ed.) *Guiness World Records, 2002.* Guiness World Records Ltd., Records Research Services, Guiness World Records, 338 Euston Road, London NWI 3BD, ISBN 1-892051-06-0, pp. 14 and 72.

Frankham, R.; Ballou, J. D. and Briscoe, D. A. (2002) *Introduction to Conservation Genetics*. Cambridge University Press, Cambridge, pp. 1-528.

Galea, P.; Scott, J. M. and Goel, K. M. (1982) Feto-fetal transfusion syndrome. *Arch. Dis. Child.* 57:781.

Galton, F. (1895) *Finger Print Directories*. Macmillan and Co., London and New York, pp. 1-123.

Galton, F. (1965) *Finger Prints*. Macmillan and Company, London, pp. 1-212.

Gardner, E. J.; Simmons, M. J. and Snustad, D. P. (1991) *Principles of Genetics*. Eighth Edition. John Wiley and Sons, Inc., pp. 1-648.

Garner, C.; Tatu, T.; Reitlie, J. E.; Littlewood, T.; Darley, J.; Cervino, S.; Farrall, M.; Kelly, P.; Spector, T. D. and Tehin, S. L. (2000) Genetic influences on F cells and other hematologic variables: a twin heritable study. *Blood,* 95(1): 342-346.

Ghose, V. (2000) *(Ed) Limca Book of Records.* Millennium Edition. The Coca Cola Company, USA, pp 1-464.

Gilbert, S. F. (2003) *Developmental Biology.* Seventh Edition. Sinauer Associates, Inc., Sunderland, Massachusetts, pp 1-784.

Goetz, Philip W. (1979) (Editior in Chief) *Encyclopedia Britannica.* 15th Edition. Encyclopedia Britannica, Inc., Chicago, Auckland, Geneva, London, Madrid, Manila, Paris, Rome, Seoul, Sydney, Tokyo and Toronto, Vol. IX, pp. 177 and Vol. X, pp. 213.

Gordon, M. S.; Bartholomew, G. A.; Grinnell, A. D.; Jorgensen, C. B. and White, F. N. (1977) *Animal Physiology. Third Edition.* Mac millan Publishing Co., Inc., New York and Collier Macmillan Publishers, London, USA, pp. 1-675.

Gottesman, I. I. and Shields, J. (1972) *Schizophrenia and Genetics : A Twin Study Vantage Point.* Academic Press, New York, pp. 1-187.

Grant, A. (1989) Cervical cerclage to prolong pregnancy. In Chalmers I, Enkin M, Keirse MJNC (eds) *Effective care in pregnancy and childbirth.* Oxford University Press, Oxford, pp. 633-645.

Gurling, H. M.; Murray, R. M. and Clifford, C. A. (1981) Investigations into the genetics of alcohol dependence and into its effects on brain functions. *Prog. Clin. Biol. Res.*, 69 (Pt C) : 77-87.

Guyton, A. C. and Hall, J. E. (2000) *Text Book of Medical Physiology.* Tenth Edition. Elsevier, a Division of Reed Elsevier India Private Limited, New Delhi, pp. 1-978.

Hamilton, W. J. (1944) Phases of maturation and fertilization in the human ova. *J. Anat.*, 78:1-4.

Hardless, H. R. (1970) *Handwriting and Thumb Print Identification and Forensic Science.* Law Book Company, Allahabad, pp. 1-291.

Hardy, G. H. (1908) Mendelian proportions in mixed population. *Science,* 28:49-50.

Harrison, W. R. (1958) *Suspect Documents: Their Scientific Examinations.* Sweet and Maxwell Limited, 2 and 3 Chancerylane, London, pp. 288-348.

Hartl, D. L. and Jones, E. W. (2001) *Genetics.* Fifth Edition. Jones and Bartlett, Sudbury, Massachusetts, pp. 1-778.

Hay, S. and Wehrung, D. (1970) Congenital malformations in twins. *Am. J. Hum. Genet.*, 22: 662-78.

Heath, A. C.; Bucholz, K. K.; Madden, P. A. F.; Dinwiddie, S. H.; Slutske, W. S,; Bierut, L. J.; Statham, D. J.; Dunne, M. P.; Whitfield, J. B. and Martin, N. G. (1997) Genetic and environmental contributions to alcohol dependence risk in a national twin sample : consistency in findings in women and men. *Psychol. Med.*, 27: 1381-1396.

Herbert, M. C. and Graham, C. F. (1974) Cell determination and biochemical differentiation of early mammalian embryo. *Curr. Top. Dev. Biol.*, 8: 151-178.

Holt, Arthur G. (1965) *Handwriting in Psychological Interpretations.* Charles C. Thomas, Springfield, Illinois, USA, pp. 1-262.

Hrubec, Z. and Omenn, G. S. (1981) Evidence of genetic predisposition to alcoholic cirrhosis and psychosis: twin concordances for alcoholism and its biological end points by zygosity among male veterans. *Alcoholism Clin. Exp. Res.*, 5: 207-215.

Jacob, F. and Monod, J. (1961) Genetic regulatory mechanisms in the synthesis of proteins. *J. Mol. Biol.*, 3:318-356.

Jennings, H. S. (1965) *Behaviour of the lower organisms*. Oxford and IBH Publishing Co., Calcutta, pp. 1-350.

Jones, J. P. and Fox, H. (1977) Syncytial knots and intervillous bridges in the human placenta: an ultra structural study. *J. Anat.*, 124: 275-286.

Kaij, L. (1960) *Alcoholism in Twins*. Almqvist and Wiksell, Stockholm, pp. 17-21.

Kaprio, J.; Koskenvuo, M.; Langinvainio, H.; Romanov, K.; Sama, S. and Rose, R. J. (1987) Genetic influences on use and abuse of alcohol : a study of 5, 638 adult Finnish twin brothers. Alcohol. *Clin. Exp. Res.*, 11 : 349-356.

Kendler, K. S.; Heath, A. C.; Neale, M. C.; Kessler, R. C. and Eaves, C. l. (1992) A population based twin study of alcoholism in women. *JAMA*, 268: 1877-1882.

Klug, W. S. and Cummings, M. R. (2003) *Concepts of Genetics.* Seventh Edition. Pearson Education (Singapore) Pte. Ltd., Indian Branch, Delhi, India, pp. 1-692.

Konberg, A. (1974) DNA Synthesis. W. H. *Freeman and Company,* San Francisco and Toppan Company, Ltd., Tokyo, Japan, pp. 1-373.

Koskenvuo, M.; Langinvainio, H.; Kaprio, J.; Lonnquist, J. and Tienari, P. (1984) Psychiatric hospitalisation in twins. *Acta. Genet, Med, Gemellol.*, 33: 321-332.

Kynaston, N. (2000) (Ed.) *Guiness World Records 2000.* Millennium Edition. Guiness World Records Ltd., Guiness Media Inc., USA, ISBN 0-85112-098-9, pp. 152-154.

Landy, H. J.; Weiner, S.; Corson, S. L.; Batzer, F. R. and Bolognese, R. J. (1986) 'The vanishing twin':Ultrasonographic assessment of fetal disappearance in the first trimester. *J. Obst. and Gynecol.* 155, 14-19.

Langman, J. (1969) *Medical Embryology.* Willams and Wilkins, Baltimore, pp. 1-185.

Larson, J. A. (1924) *Single Finger Print System.* D. Appleton and Company, New York, USA, pp. 1-249.

Lindemann, J. P.; Kang, K. W. and Christian, J. C. (1977) Genetic variance of erythrocyte parameters in adult male twins. *Clin. Genet.,* 12(2) : 73-76.

Lewin, B. (2002) *Genes VII.* Oxford University Press, Oxford, pp. 1-952.

Mahone, P. R.; Sherer, D. M.; Abramowicz, J. S. and Woods, J. R. (1993) Twin-twin transfusion syndrome : Rapid development of severe hydrops of the donor following selective feticide of the hydropic recipient. *Am. J. Obstet. Gynecol.*, 169 : 166.

Markert, C. L. and Ursprung, H. (1971) *Developmental Genetics.* Prentice-Hall, Inc., Engle Wood Cliffs, N. J., USA, pp. 1-211.

Mather, P. L. and Black, K. N. (1984) Heredity and environmental influences on preschool twins' language skills. *Devl. Psychol.,* 20: 303-308.

Mc Clintock, B. (1941) The association of mutants with homozygous deficiencies in *Zea mays*. Genetics, 26: 542-571.

Mc Clintock, B. (1944) The relation of homozygous deficiencies to mutations and allelic series in maize. *Genetics*, 29: 478-502.

McGue, M.; Pickens, R. W. and Svikis, D. S. (1992) Sex and age effects on the inheritance of alcohol problems : a twin study. *J. Abnormal Psychology*, 101: 3-17.

Mc Lachlam, J. (1994) *Medical Embryology.* Addison Wesley Publishing Company, Wokingham, England, pp. 62-68.

Mehta, M. K. (1970) *Indentification of Handwriting and Cross Examination of Experts.* N. M. Tripathi Private Ltd., pp. 1-399.

Meyers, S. M. (1994) A twin study on age-related macular degeneration. Trans. *Am. Ophthalmol. Soc.*, 92 : 775-843.

Meyers, S. M.; Greene, T. and Gutman, F. A. (1995) A twin study of age-related macular degeneration. *Am. J. Ophthalmol.*, 120: 757-766.

Meyers, S. M. and Zachary, A. A. (1988) Monozygotic twins with age-related macular degeneration. *Arch. Ophthalmol.*, 106: 651-653.

Micklos, D. A.; Freyer, G. A. and Crotty, D. A. (2003) DNA Science: A First Course. Second Edition. Cold Spring Harbor Laboratory Press, Cold Spring Harbor, New York, pp. 1- 499.

Mittler, P. (1976) Language development in young twins : biological, genetic and social aspects. *Acta Genet. Med. Gemell.* 25 : 359-365.

Moenssens, A. A. (1975) *Finger Print Techniques*. First Edition. Chilton Book Company, Pennsylvania, pp. 1-301.

Morton, N. E. (1972) Human behavioral genetics. In *Genetics, Environment and Behavior*. Academic Press, New York, pp. 1-257.

Muller, H. J. (1974) *Studies in Genetics*. Oxford and IBH Publishing Co. New Delhi, pp. 1-588.

Muller, H. J. and Altenburg, E. (1919) The rate of change of hereditary factors in *Drosophila*. *Proc. Sac. Exper. Biol. and Med.*, 17:10-14.

Myrianthopoulos, N. (1970) An epidemiologic survey of twins in a large prospectively studied population. *Am. J. Hum. Genet.*, 22 : 611-629.

Newman, H. H.; Freeman, F. N. and Holzinger, K. J. (1937) *Twins : a Study of Heredity and Environment.* University of Chicago Press, Chicago, pp. 1-308.

Newman, R. B.;Hamer, C. and Miller, M. C. (1989) Outpatient triplet management: a contemporary review. *Am. J. Obstet. Gynecol.*, 161:547-555.

Nolan, F. and Oh, T. (1996) Identical twins, different voices. *Forensic Ling*, 3:39-49.

Novitski, Edward (1977) *Human Genetics*. Macmillan Publising Co., Inc. New York and Collier Mac millan Publishers, London, pp. 1-458.

Osborn, Albert S. (1929) *Questioned Documents.* Boyd Printing Co. Albany, Toronto, pp. 1-1042.

Panneerselvam, R. (2004) *Research Methodology,* Prentice-Hall of India Private Limited, New Delhi, pp. 1- 598.

Pickens, R. W.; Svikis, D. S.; McGue, M.; Lykken, D. T.; Heston, L. L. and Clayton, P. J. (1991) Heterogeneity in the inheritance of alcoholism : A study of male and female twins. *Arch. Gen Psychiatry,* 48: 19-28.

Pinkerton, J. H. M. (1961) Development of the human ovary – a study using histochemical techniques. *Obstat. Gynec.*, 18: 152-181.

Prescott, C. A. and Kendler, K. S. (1999) Genetic and environmental contributions to alcohol abuse and dependence in a population-based sample of male twins. *Am. J. Psychiatry, 156* : 34-40.

Przybyla, B. D.; Horii Y. and Crawford, M. H. (1992) Vocal fundamental frequency in a twin sample: looking for a genetic effect. *J. Voice,* 3: 261-266.

Rao, P. S. S. S and Richard, J. (2004) *An Introduction to Biostatistics*. Third Edition. Prentice-Hall of India Private Limited, New Delhi, India, pp. 1-233.

Raven, J. C. (1962) *Coloured Progressive Matrices.* H. K. Lewis and Co. Ltd., London, pp. A_1-B_{12}.

Richard, Landsdown (1985) *Child Development: Made Simple.* Rupa and Co. by arrangement with William Heinemann Ltd., London, pp. 1-256.

Robertson, E. G. and Neer, K. J. (1988) Placental injection studies in twin gestation. *Am. J. Obstet. Gynecol.*, 147:170.

Romanov, K.; Kaprio, J.; Rose, R. and Koskenvuo, M. (1991) Genetics of alcoholism: effects of migration on concordance rates among male twins. *Alcohol alcoholism (Suppl.) 1*: 137-140.

Rotham, D. (1970) Folic acid in pregnancy. *Am. J. Obstet. Gynecol.,* 108:149-175.

Saudek, Robert (1954) *The Psychology of Handwriting*. George Allen and Unwin Ltd., Ruskin House, Museum Street, London, pp. 1-288.

Schmidt-Nelson, K. (1977) *Animal Physiology.* Fifth Edition. Cambridge University Press, Cambridge, pp. 1-574.

Scott, W. R. (1951) *Finger Print Mechanics.* Charles C Thomas Publisher, Springfield, Illinois, USA. pp. 1-432.

Segal, N. L. (1990) The importance of twin studies for individual differences research. *Journal of Counselling and Development,* 68: 612-622.

Segal, N. L. (1993) Twin, sibling, and adoption methods. *American Psychologist,* 48(9): 943-956.

Segal, N. L. (1997a) Behavioural aspects of intergenerational human cloning: What twins tell us. *Jurimetrics*, 38: 57-67.

Segal, N. L. (1997b) Same age unrelated siblings: a unique test of within family environmental influences on IQ similarity. *Journal of Educational Psychology,* 89(2):381-390.

Segal, N. L. and Hershberger, S. L. (1999) Cooperation and competition between twins: findings from a prisoner's dilemma game. *Evolution and Human Behaviour,* 20 : 29-51.

Segal, N. L. and Mac Donald, K. B. (1998) Behavioural genetics and evolutionary psychology : unified perspective on personality research. *Human Biology,* 70(2): 159-184.

Segal, N. L. and Ream, S. L. (1998) Decrease in grief intensity for deceased twin and non-twin relatives: an evolutionary perspective. *Personality and Individual Differences,* 25:317-325.

Segal, N. L. and Russell, J. M. (1992) Twins in the class room: school policy issues and recommendations. *Journal of Educational and Psychological Consultation,* 3(1):69-84.

Segal, N. L.; Topolski, T. D.; Wilson, S. M.; Brown, K. W. and Araki, L. (1994) Twin analysis of odor identification and perception. *Physiology and Behaviour,* 57(3): 605-609.

Shettles, L. B. (1958) The living human ovum. *Am. J. Osbtat. Gynec.,* 76:398-406.

Shields, J. (162) *Monozygotic Twins Brought up Apart and Brought up Together.* Oxford University Press, London, pp. 1-219.

Singer, E. (1969) *A Manual of Graphology.* Un win Brothers Limited, The Gresham Press, Great Britain, pp. 1-244.

Snustad, D. P. and Simmons, M. J. (2003) *Principles of Genetics.* John Wiley and Sons Inc., New York, pp. 1-757.

Spellacy, W. N. Multiple Pregnancies. In Scott, J. R; DiSAIA P. J.; Hammond, C. B. and Spellacy, W. N. (eds)(1994) *Danforth's Obstetrics and Gynecology,* seventh edition, J. B. Lippincott Company, Philadelphia, pp. 333-340.

Stassen, H. H.; Lykken, D. T.; Propping, P. and Bomben, G. (1988) Genetic determination of the human EEG. Survey of recent results on twins reared together and apart. *Hum. Genet.,* 80 : 165-176.

Strickberger, M. W. (1985) *Genetics.* Third Edition. Prentice-Hall, Inc., USA, pp. 1-782.

Sturtevant, A. H. (1965) *A History of Genetics.* Harper and Row Publishers, Inc., New York, pp. 1-135.

Teltscher, Henry, O. (1948) *Handwriting : The key to Successful Living.* G. P. Putnam's Sons, New York, pp. 1-313.

Tienthoven, V. (1968) *Reproductive Physiology of Vertebrates.* W. B. Saunders Company, Philadelphia, London and Toronto, pp. 1-473.

Tucker, W. H. (1997) Re-reconsidering burt: beyond a reasonable doubt. *Journal of the History of the Behavioural Sciences,* 33 (2): 145-162.

Turpin, R. and Lejeune, J. (1969) Monozygotic twinning and chromosome aberrations (Heterokaryotic monozygotism). In R. Turpin and J. Lejeune, eds., *Human Affliction and Chromosomal Aberrations.* Pergamon, Oxford, pp. 1-407.

Unnithan, S. and Ghosh, I. (2005) *Twin Hopes.* India Today. Living media India Ltd., New Delhi, 30 (41): 56-64.

Ursprung, H. (1967) The Formation of Patterns in Development. In M. Locke (Ed.) *Major problems in developmental biology.* Academic Press, New York, pp. 177-216.

Vogel, F.; Schalt, E.; Kruger, J. and Klarish, G. (1982) Relationship between behavioural maturation measured by the "Baum" that and EEG frequency. A pilot study on monozygotic and dizygotic twins. *Hum. Genet.*, 62: 60-65.

Wagner, R. P. (1955) *Genetics and Metabolism.* John Wiley and Sons, Inc., New York and Chapman and Hall, Limited, London, pp. 1-411.

Watson, J. D. and Crick, F. H. C. (1953 a) Genetical implications of the structure of deoxyribonucleic acid. *Nature*, 171:964-967.

Watson, J. D. and Crick, F. H. C. (1953 b) Molecular structure of nucleic acids: A structure of deoxyribose nucleic acid. *Nature*, 171:737-738.

Weil, A. J. (1965) The spermatozoa-coating antigen (S. C. A.) of the seminal vesicle. *Ann. NY Acad. Sci.*, 124:267-269.

Whitfield, J. B. and Martin, N. G. (1985) Genetic and environmental influences in the size and number of cells in the blood. *Genet. Epidemiol.*, 2(2): 133-144.

Whiteside, S. P. and Rixon, E. (2001) Speech patterns of monozygotic twins: an acoustic case study of monosyllabic words. *Phonetician,* 84: 9-22.

Wolpert, L. (2002) *Principles of Development.* Second Edition. Oxford University Press Inc., New York, USA, pp. 1-518.

Yokoyama, Y. and Akiyama, T. (1995) Intrapair differences of the blood cell components and lymphocyte subsets in monozygotic and dizygotic twins. *Acta. Genet. Med. Gemellol. (Roma)*, 44 (3-4): 203-214.

Young, J. P.; Lader, M. H. and Fenton, G. W. (1972) A twin study on the genetic influences on the electroencephalogram. *J. Med. Genet.,* 9:13-16.

Zar, J. H. (2003) *Biostatistical Analysis.* Fourth Edition. Pearson Education (Singapore) Pte. Ltd., Indian Branch, Delhi, India, pp. 1-663.

Zhu, G.; Duffy, D. L.; Eldridge, A.; Grace, M.; Mayne, C.; O'Gorman, L.; Aitken, J. F.; Neale, M. C.; Hayward, N. K.; Green, A. C. and Mortin, N. G. (1999) A major quantitative trait locus for mole density is linked to the familial melanoma gene CDKN2A: a maximum-likelihood combined linkage and association analysis in twins and their sibs. Am. *J. Hum. Genet.*, 65 (2) : 483-492.

Zuckerman, P. L. and Weir, B. J. (1977) *The Ovary*. Second Edition. Academic Press Inc., New York, pp. 1-467.